Computer Architecture Performance Evaluation Methods

Computer Architecture Performance Evaluation Methods

Lieven Eeckhout

ISBN: 978-3-031-00599-2   paperback
ISBN: 978-3-031-01727-8   ebook

DOI 10.1007/978-3-031-01727-8

A Publication in the Springer series
SYNTHESIS LECTURES ON ADVANCES IN AUTOMOTIVE TECHNOLOGY

Lecture #10
Series Editor: Mark D. Hill, *University of Wisconsin*
Series ISSN
Synthesis Lectures on Computer Architecture
Print 1935-3235   Electronic 1935-3243

Synthesis Lectures on Computer Architecture

Editor
Mark D. Hill, *University of Wisconsin*

Synthesis Lectures on Computer Architecture publishes 50- to 100-page publications on topics pertaining to the science and art of designing, analyzing, selecting and interconnecting hardware components to create computers that meet functional, performance and cost goals. The scope will largely follow the purview of premier computer architecture conferences, such as ISCA, HPCA, MICRO, and ASPLOS.

Chip Multiprocessor Architecture: Techniques to Improve Throughput and Latency
Kunle Olukotun, Lance Hammond, and James Laudon
2007

Transactional Memory
James R. Larus and Ravi Rajwar
2006

Quantum Computing for Computer Architects
Tzvetan S. Metodi and Frederic T. Chong
2006

Computer Architecture Performance Evaluation Methods

Lieven Eeckhout
Ghent University

SYNTHESIS LECTURES ON COMPUTER ARCHITECTURE #10

ABSTRACT

Performance evaluation is at the foundation of computer architecture research and development. Contemporary microprocessors are so complex that architects cannot design systems based on intuition and simple models only. Adequate performance evaluation methods are absolutely crucial to steer the research and development process in the right direction. However, rigorous performance evaluation is non-trivial as there are multiple aspects to performance evaluation, such as picking workloads, selecting an appropriate modeling or simulation approach, running the model and interpreting the results using meaningful metrics. Each of these aspects is equally important and a performance evaluation method that lacks rigor in any of these crucial aspects may lead to inaccurate performance data and may drive research and development in a wrong direction.

    The goal of this book is to present an overview of the current state-of-the-art in computer architecture performance evaluation, with a special emphasis on methods for exploring processor architectures. The book focuses on fundamental concepts and ideas for obtaining accurate performance data. The book covers various topics in performance evaluation, ranging from performance metrics, to workload selection, to various modeling approaches including mechanistic and empirical modeling. And because simulation is by far the most prevalent modeling technique, more than half the book's content is devoted to simulation. The book provides an overview of the simulation techniques in the computer designer's toolbox, followed by various simulation acceleration techniques including sampled simulation, statistical simulation, parallel simulation and hardware-accelerated simulation.

KEYWORDS

computer architecture, performance evaluation, performance metrics, workload characterization, analytical modeling, architectural simulation, sampled simulation, statistical simulation, parallel simulation, FPGA-accelerated simulation

Contents

Preface

GOAL OF THE BOOK

The goal of this book to present an overview of the current state-of-the-art in computer architecture performance evaluation, with a special emphasis on methods for exploring processor architectures. The book focuses on fundamental concepts and ideas for obtaining accurate performance data. The book covers various aspects that relate to performance evaluation, ranging from performance metrics, to workload selection, and then to various modeling approaches such as analytical modeling and simulation. And because simulation is, by far, the most prevalent modeling technique in computer architecture evaluation, more than half the book's content is devoted to simulation. The book provides an overview of the various simulation techniques in the computer designer's toolbox, followed by various simulation acceleration techniques such as sampled simulation, statistical simulation, parallel simulation, and hardware-accelerated simulation.

The evaluation methods described in this book have a primary focus on performance. Although performance remains to be a key design target, it no longer is the sole design target. Power consumption, energy-efficiency, and reliability have quickly become primary design concerns, and today they probably are as important as performance; other design constraints relate to cost, thermal issues, yield, etc. This book focuses on performance evaluation methods only, and while many techniques presented here also apply to power, energy and reliability modeling, it is outside the scope of this book to address them. This does not compromise on the importance and general applicability of the techniques described in this book because power, energy and reliability models are typically integrated into existing performance models.

The format of the book is such that it cannot present all the details of all the techniques described in the book. As mentioned before, the goal is to present a broad overview of the state-of-the-art in computer architecture performance evaluation methods, with a special emphasis on general concepts. We refer the interested reader to the bibliography for in-depth treatments of the specific topics covered in the book, including three books on performance evaluation [92; 94; 126].

BOOK ORGANIZATION

This book is organized as follows, see also Figure 1.

Chapter 2 describes ways to quantify performance and revisits performance metrics for single-threaded workloads, multi-threaded workloads and multi-program workloads. Whereas quantifying performance for single-threaded workloads is straightforward and well understood, some may still be confused about how to quantify multi-threaded workload performance. This is especially true for

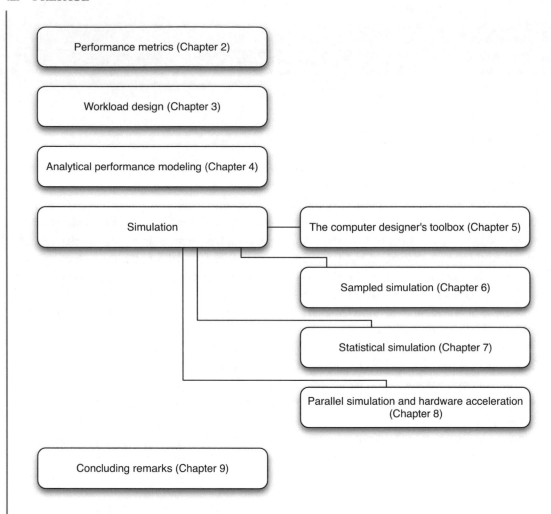

Figure 1: Book's organization.

quantifying multiprogram performance. This book sheds light on how to do a meaningful multiprogram performance characterization by focusing on both system throughput and job turnaround time. We also discuss ways for computing the average performance number across a set of benchmarks and clarify the opposite views on computing averages, which fueled the debate over the past two decades.

Chapter 3 talks about how to select a representative set of benchmarks from a larger set of specified benchmarks. The chapter covers two methodologies for doing so, namely Principal Component Analysis and the Plackett and Burman design of experiment. The idea behind both methodologies is that benchmarks that exhibit similar behavior in their inherent behavior and/or their interaction with the microarchitecture should not both be part of the benchmark suite, only dissimilar benchmarks should. By retaining only the dissimilar benchmarks, one can reduce the number of retained benchmarks and thus reduce overall experimentation time while not sacrificing accuracy too much.

Analytical performance modeling is the topic of Chapter 4. Although simulation is the prevalent computer architecture performance evaluation method, analytical modeling clearly has its place in the architect's toolbox. Analytical models are typically simple and, therefore, very fast. This allows for using analytical models to quickly explore large design spaces and narrow down on a region of interest, which can later be explored in more detail through simulation. Moreover, analytical models can provide valuable insight, which is harder and more time-consuming to obtain through simulation. We discuss three major flavors of analytical models. Mechanistic modeling or white-box modeling builds a model based on first principles, along with a good understanding of the system under study. Empirical modeling or black-box modeling builds a model through training based simulation results; a model, typically, is a regression model or a neural network. Finally, hybrid mechanistic-empirical modeling aims at combining the best of worlds: it provides insight (which it inherits from mechanistic modeling) while easing model construction (which it inherits from empirical modeling).

Chapter 5 gives an overview of the computer designer's toolbox while focusing on simulation methods. We revisit different flavors of simulation, ranging from functional simulation, (specialized) trace-driven simulation, execution-driven simulation, full-system simulation, to modular simulation infrastructures. We describe a taxonomy of execution-driven simulation, and we detail on ways for how to deal with non-determinism during simulation.

The next three Chapters 6, 7, and 8, cover three approaches to accelerate simulation, namely sampled simulation, statistical simulation and through exploiting parallelism. Sampled simulation (Chapter 6) simulates only a small fraction of a program's execution. This is done by selecting a number of so called sampling units and only simulating those sampling units. There are three challenges for sampled simulation: (i) what sampling units to select; (ii) how to initialize a sampling unit's architecture starting image (register and memory state); (iii) how to estimate a sampling unit's microarchitecture starting image, i.e., the state of the caches, branch predictor, and processor core at the beginning of the sampling unit. Sampled simulation has been an active area of research over the past few decades, and this chapter covers the most significant problems and solutions.

Statistical simulation (Chapter 7) takes a different approach. It first profiles a program execution and collects some program metrics that characterize the program's execution behavior in a statistical way. A synthetic workload is generated from this profile; by construction, the synthetic workload exhibits the same characteristics as the original program. Simulating the synthetic workload on a simple statistical simulator then yields performance estimates for the original workload. The key benefit is that the synthetic workload is much shorter than the real workload, and as a result, simulation is done quickly. Statistical simulation is not meant to be a replacement for detailed cycle-accurate simulation but rather as a useful complement to quickly explore a large design space.

Chapter 8 covers three ways to accelerate simulation by exploiting parallelism. The first approach leverages multiple machines in a simulation cluster to simulate multiple fragments of the entire program execution in a distributed way. The simulator itself may still be a single-threaded program. The second approach is to parallelize the simulator itself in order to leverage the available parallelism in existing computer systems, e.g., multicore processors. A parallelized simulator typically exploits coarse-grain parallelism in the target machine to efficiently distribute the simulation work across multiple threads that run in parallel on the host machine. The third approach aims at exploiting fine-grain parallelism by mapping (parts of) the simulator on reconfigurable hardware, e.g., Field Programmable Gate Arrays (FPGAs).

Finally, in Chapter 9, we briefly discuss topics that were not (yet) covered in the book, namely measurement bias, design space exploration and simulator validation, and we look forward towards the challenges ahead of us in computer performance evaluation.

Lieven Eeckhout
June 2010

Acknowledgments

First and foremost, I would like to thank Mark Hill and Michael Morgan for having invited me to write a synthesis lecture on computer architecture performance evaluation methods. I was really honored when I received the invitation from Mark, and I really enjoyed working on the book.

Special thanks also to my reviewers who have read early versions of the book and who gave me valuable feedback, which greatly helped me improve the text. Many thanks to: Joel Emer, Lizy John, Babak Falsafi, Jim Smith, Mark Hill, Brad Calder, Benjamin Lee, David Brooks, Amer Diwan, Joshua Yi and Olivier Temam.

I'm also indebted to my collaborators over the past years who have given me the opportunity to learn more and more about computer architecture in general and performance evaluation methods in particular. This book comprises many of their contributions.

Last but not least, I would like to thank my wife, Hannelore Van der Beken, for her endless support throughout this process, and our kids, Zoë, Jules, Lea and Jeanne for supporting me — indirectly — through their love and laughter.

Lieven Eeckhout
June, 2010

CHAPTER 1

Introduction

Performance evaluation is at the foundation of computer architecture research and development. Contemporary microprocessors are so complex that architects cannot design systems based on intuition and simple models only. Adequate performance evaluation methodologies are absolutely crucial to steer the development and research process in the right direction. In order to illustrate the importance of performance evaluation in computer architecture research and development, let's take a closer look at how the field of computer architecture makes progress.

1.1 STRUCTURE OF COMPUTER ARCHITECTURE (R)EVOLUTION

Joel Emer in his Eckhert-Mauchly award speech [56] made an enlightening analogy between scientific research and engineering research (in this case, computer systems research) [55]. The scientific research side of this analogy is derived from Thomas Kuhn's theory on the evolution of science [110] and describes scientific research as is typically done in five steps; see Figure 1.1(a). The scientist first takes a hypothesis about the environment and, subsequently, designs an experiment to validate the hypothesis — designing an experiment often means taking a random sample from a population. The experiment is then run, and measurements are obtained. The measurements are then interpreted, and if the results agree with the hypothesis, additional refined experiments may be done to increase confidence in the hypothesis (inner loop in Figure 1.1); eventually, the hypothesis is accepted. If the outcome of the experiment disagrees with the hypothesis, then a new hypothesis is needed (outer loop), and this may potentially lead to what Kuhn calls a scientific revolution.

The procedure in systems research, in general, and computer architecture research and development, in particular, shows some similarities, see Figure 1.1(b). The architect takes a hypothesis about the environment — this is typically done based on intuition, insight and/or experience. One example hypothesis may be that integrating many small in-order processor cores may yield better overall system throughput than integrating just a few aggressive out-of-order processor cores on a single chip. Subsequently, the architect designs an experiment to validate this hypothesis. This means that the architect picks a baseline design, e.g., the architect determines the processor architecture configurations for the in-order and out-of-order architectures; in addition, the architects picks a number of workloads, e.g., the architect collects a number of compute-intensive, memory-intensive and/or I/O-intensive benchmarks. The experiment is then run, and a number of measurements are obtained. This can be done by running a model (analytical model or simulation model) and/or by doing a real hardware experiment. Architects typically run an extensive number of measurements

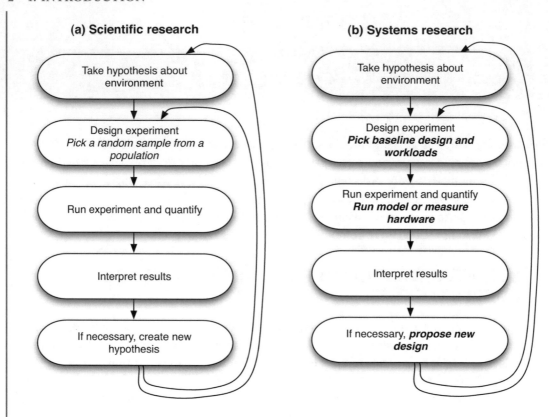

Figure 1.1: Structure of (a) scientific research versus (b) systems research.

while sweeping through the design space. The key question then is to navigate through this wealth of data and make meaningful conclusions. Interpretation of the results is crucial to make correct design decisions. The insights obtained from the experiment may or may not support the hypothesis made by the architect. If the experiment supports the hypothesis, the experimental design is improved and additional experimentation is done, i.e., the design is incrementally refined (inner loop). If the experiment does not support the hypothesis, i.e., the results are completely surprising, then the architect needs to re-examine the hypothesis (outer loop), which may lead the architect to change the design or propose a new design.

Although there are clear similarities, there are also important differences that separate systems research from scientific research. Out of practical necessity the step of picking a baseline design and workloads is typically based on the experimenters judgment and experience, rather than objectively drawing a scientific sample from a given population. This means that the architect should be well aware of the subjective human aspect of experiment design when interpreting and analyzing the results. Trusting the results produced through the experiment without a clear understanding of its design may lead to misleading or incorrect conclusions. In this book, we will focus on the scientific approach suggested by Kuhn, but we will also pay attention to making the important task of workload selection less subjective.

1.2   IMPORTANCE OF PERFORMANCE EVALUATION

The structure of computer architecture evolution, as described above, clearly illustrates that performance evaluation is at the crux of computer architecture research and development. Rigorous performance evaluation is absolutely crucial in order to make correct design decisions and drive research in a fruitful direction. And this is even more so for systems research than is the case for scientific research: as argued above, in order to make meaningful and valid conclusions, it is absolutely crucial to clearly understand how the experimental design is set up and how this may affect the results.

The structure of computer architecture evolution also illustrates that there are multiple aspects to performance evaluation, ranging from picking workloads, picking a baseline design, picking an appropriate modeling approach, running the model, and interpreting the results in a meaningful way. And each of these aspects is equally important — a performance evaluation methodology that lacks rigor in one of these crucial aspects may fall short. For example, picking a representative set of benchmarks, picking the appropriate modeling approach, and running the experiments in a rigorous way may still be misleading if inappropriate performance metrics are used to quantify the benefit of one design point relative to another. Similarly, inadequate modeling, e.g., a simulator that models the architecture at too high a level of abstraction, may either underestimate or overestimate the performance impact of a design decision, even while using the appropriate performance metrics and benchmarks.

Architects are generally well aware of the importance of adequate performance evaluation, and, therefore, they pay detailed attention to the experimental setup when evaluating research ideas and design innovations.

1.3   BOOK OUTLINE

This book presents an overview of the current state-of-the-art in computer architecture performance evaluation, with a special emphasis on methods for exploring processor architectures. The book covers performance metrics (Chapter 2), workload design (Chapter 3), analytical performance modeling (Chapter 4), architectural simulation (Chapter 5), sampled simulation (Chapter 6), statistical simulation (Chapter 7), parallel simulation and hardware acceleration (Chapter 8). Finally, we conclude in Chapter 9.

CHAPTER 2

Performance Metrics

Performance metrics are at the foundation of experimental research and development. When evaluating a new design feature or a novel research idea, the need for adequate performance metrics is paramount. Inadequate metrics may be misleading and may lead to incorrect conclusions and may steer development and research in the wrong direction. This chapter discusses metrics for evaluating computer system performance. This is done in a number of steps. We consider metrics for single-threaded workloads, multi-threaded workloads and multi-program workloads. Subsequently, we will discuss ways of summarizing performance in a single number by averaging across multiple benchmarks. Finally, we will briefly discuss the utility of partial metrics.

2.1   SINGLE-THREADED WORKLOADS

Benchmarking computer systems running single-threaded benchmarks is well understood. The total time T to execute a single-threaded program is the appropriate metric. For a single-threaded program that involves completing N instructions, the total execution time can be expressed using the 'Iron Law of Performance' [169]:

$$T = N \times CPI \times \frac{1}{f}, \qquad (2.1)$$

with CPI the average number of cycles per useful instruction and f the processor's clock frequency. Note the wording 'useful instruction'. This is to exclude the instructions executed along mispredicted paths — contemporary processors employ branch prediction and speculative execution, and in case of a branch misprediction, speculatively executed instructions are squashed from the processor pipeline and should, therefore, not be accounted for as they don't contribute to the amount of work done.

The utility of the Iron Law of Performance is that the terms correspond to the sources of performance. The number of instructions N is a function of the instruction-set architecture (ISA) and compiler; CPI is a function of the micro-architecture and circuit-level implementation; and f is a function of circuit-level implementation and technology. Improving one of these three sources of performance improves overall performance. Justin Rattner [161], in his PACT 2001 keynote presentation, reported that x86 processor performance has improved by over $75\times$ over a 10 years time period, between the 1.0μ technology node (early 1990s) and the 0.18μ technology node (around 2000): $13\times$ comes from improvements in frequency, and $6\times$ from micro-architecture enhancements, and the $50\times$ improvement in frequency in its turn is due to improvements in technology ($13\times$) and micro-architecture ($4\times$).

Assuming that the amount of work that needs to be done is constant, i.e., the number of dynamically executed instructions N is fixed, and the processor clock frequency f is constant, one can express the performance of a processor in terms of the CPI that it achieves. The lower the CPI, the lower the total execution time, the higher performance. Computer architects frequently use its reciprocal, or IPC, the average number of (useful) instructions executed per cycle. IPC is a higher-is-better metric. The reason why IPC (and CPI) are popular performance metrics is that they are easily quantified through architectural simulation. Assuming that the clock frequency does not change across design alternatives, one can compare microarchitectures based on IPC.

Although IPC seems to be more widely used than CPI in architecture studies — presumably, because it is a higher-is-better metric — CPI provides more insight. CPI is additive, and one can break up the overall CPI in so called CPI adders [60] and display the CPI adders in a stacked bar called the CPI stack. The base CPI is typically shown at the bottom of the CPI stack and represents useful work done. The other CPI adders, which reflect 'lost' cycle opportunities due to miss events such as branch mispredictions and cache and TLB misses, are stacked on top of each other.

2.2   MULTI-THREADED WORKLOADS

Benchmarking computer systems that run multi-threaded workloads is in essence similar to what we described above for single-threaded workloads and is adequately done by measuring the time it takes to execute a multi-threaded program, or alternatively, to get a given amount of work done (e.g., a number of transactions for a database workload), assuming there are no co-executing programs.

In contrast to what is the case for single-threaded workloads IPC is not an accurate and reliable performance metric for multi-threaded workloads and may lead to misleading or incorrect conclusions [2]. The fundamental reason is that small timing variations may lead to different execution paths and thread interleavings, i.e., the order in which parallel threads enter a critical section may vary from run to run and may be different across different microarchitectures. For example, different interrupt timing may cause system software to take different scheduling decisions. Also, different microarchitectures may lead to threads reaching the critical sections in a different order because threads' performance may be affected differently by the microarchitecture. As a result of these differences in timing and thread interleavings, the total number of instructions executed may be different across different runs, and hence, the IPC may be different. For example, threads executing spin-lock loop instructions before acquiring a lock increase the number of dynamically executed instructions (and thus IPC); however, these spin-lock loop instructions do not contribute to overall execution time, i.e., the spin-lock loop instructions do not represent useful work. Moreover, the number of spin-lock loop instructions executed may vary across microarchitectures. Spin-lock loop instructions is just one source of error from using IPC. Another source relates to instructions executed in the operating system, such as instructions handling Translation Lookaside Buffer (TLB) misses, and idle loop instructions.

Alameldeen and Wood [2] present several case studies showing that an increase in IPC does not necessarily imply better performance, or a decrease in IPC does not necessarily reflect performance loss for multi-threaded programs. They found this effect to increase with an increasing number of processor cores because more threads are spending time in spin lock loops. They also found this effect to be severe for workloads that spend a significant amount of time in the operating system, e.g., commercial workloads such as web servers, database servers, and mail servers.

Wenisch et al. [190] consider user-mode instructions only and quantify performance in terms of user-mode IPC (U-IPC). They found U-IPC to correlate well with the number of transactions completed per unit of time for their database and web server benchmarks. The intuition is that when applications do not make forward progress, they often yield to the operating system (e.g., the OS idle loop or spin-lock loops in OS code). A limitation for U-IPC is that it does not capture performance of system-level code, which may account for a significant fraction of the total execution time, especially in commercial workloads.

Emer and Clark [59] addressed the idle loop problem in their VAX-11/780 performance study by excluding the VMS Null process from their per-instruction statistics. Likewise, one could measure out spin-lock loop instructions; this may work for some workloads, e.g., scientific workloads where most of the spinning happens in user code at places that can be identified a priori.

2.3   MULTIPROGRAM WORKLOADS

The proliferation of multi-threaded and multicore processors in the last decade has urged the need for adequate performance metrics for multiprogram workloads. Not only do multi-threaded and multicore processors execute multi-threaded workloads, they also execute multiple independent programs concurrently. For example, a simultaneous multi-threading (SMT) processor [182] may co-execute multiple independent jobs on a single processor core. Likewise, a multicore processor or a chip-multiprocessor [151] may co-execute multiple jobs, with each job running on a separate core. A chip-multithreading processor may co-execute multiple jobs across different cores and hardware threads per core, e.g., Intel Core i7, IBM POWER7, Sun Niagara [108].

As the number of cores per chip increases exponentially, according to Moore's law, it is to be expected that more and more multiprogram workloads will run on future hardware. This is true across the entire compute range. Users browse, access email, and process messages and calls on their cell phones while listening to music. At the other end of the spectrum, servers and datacenters leverage multi-threaded and multicore processors to achieve greater consolidation.

The fundamental problem that multiprogram workloads impose to performance evaluation and analysis is that the independent co-executing programs affect each other's performance. The amount of performance interaction depends on the amount of resource sharing. A multicore processor typically shares the last-level cache across the cores as well as the on-chip interconnection network and the off-chip bandwidth to memory. Chandra et al. [24] present a simulation-based experiment that shows that the performance of individual programs can be affected by as much as 65% due to resource sharing in the memory hierarchy of a multicore processor when co-executing two independent programs. Tuck and Tullsen [181] present performance data measured on the Intel

Pentium 4 processor which is an SMT processor[1] with two hardware threads. They report that for some programs per-program performance can be as low as 71% of the per-program performance observed when run in isolation, whereas for other programs, per-program performance may be comparable (within 98%) to isolated execution.

Eyerman and Eeckhout [61] take a top-down approach to come up with performance metrics for multiprogram workloads, namely system throughput (STP) and average normalized turnaround time (ANTT). They start from the observation that there are two major perspectives to multiprogram performance: a user's perspective and a system's perspective. A user's perspective cares about the turnaround time for an individual job or the time it takes between submitting the job and its completion. A system's perspective cares about the overall system throughput or the number of jobs completed per unit of time. Of course, both perspectives are not independent of each other. If one optimizes for job turnaround time, one will likely also improve system throughput. Similarly, improving a system's throughput will also likely improve a job's turnaround time. However, there are cases where optimizing for one perspective may adversely impact the other perspective. For example, optimizing system throughput by prioritizing short-running jobs over long-running jobs will have a detrimental impact on job turnaround time, and it may even lead to starvation of long-running jobs.

We now discuss the STP and ANTT metrics in more detail, followed by a discussion on how they compare against prevalent metrics.

2.3.1   SYSTEM THROUGHPUT

We, first, define a program's normalized progress as

$$NP_i = \frac{T_i^{SP}}{T_i^{MP}},\tag{2.2}$$

with T_i^{SP} and T_i^{MP}, the execution time under single-program mode (i.e., the program runs in isolation) and multiprogram execution (i.e., the program co-runs with other programs), respectively. Given that a program runs slower under multiprogram execution, normalized progress is a value smaller than one. The intuitive understanding of normalized progress is that it represents a program's progress during multiprogram execution. For example, an NP of 0.7 means that a program makes 7 milliseconds of single-program progress during a 10 millisecond time slice of multiprogram execution.

System throughput (STP) is then defined as the sum of the normalized progress rates across all jobs in the multiprogram job mix:

$$STP = \sum_{i=1}^{n} NP_i = \sum_{i=1}^{n} \frac{T_i^{SP}}{T_i^{MP}},\tag{2.3}$$

In other words, system throughput is the accumulated progress across all jobs, and thus it is a higher-is-better metric. For example, running two programs on a multi-threaded or multicore processor may

---

[1]Intel uses the term HyperThreading rather than SMT.

cause one program to make 0.75 normalized progress and the other 0.5; the total system throughput then equals $0.75 + 0.5 = 1.25$. STP is typically larger than one: latency hiding increases system utilization and throughput, which is the motivation for multi-threaded and multicore processing in the first place, i.e., cache miss latencies from one thread are hidden by computation from other threads, or memory access latencies are hidden through memory-level parallelism (MLP), etc. Nevertheless, for some combinations of programs, severe resource sharing may result in an STP smaller than one. For example, co-executing memory-intensive workloads may evict each other's working sets from the shared last-level cache resulting in a huge increase in the number of conflict misses, and hence they have detrimental effects on overall performance. An STP smaller than one means that better performance would be achieved by running all programs back-to-back through time sharing rather than through co-execution.

2.3.2 AVERAGE NORMALIZED TURNAROUND TIME

To define the average normalized turnaround time, we first define a program's normalized turnaround time as

$$NTT_i = \frac{T_i^{MP}}{T_i^{SP}}. \tag{2.4}$$

Normalized turnaround time quantifies the user-perceived slowdown during multiprogram execution relative to single-program execution, and typically is a value larger than one. NTT is the reciprocal of NP.

The average normalized turnaround time is defined as the arithmetic average across the programs' normalized turnaround times:

$$ANTT = \frac{1}{n} \sum_{i=1}^{n} NTT_i = \frac{1}{n} \sum_{i=1}^{n} \frac{T_i^{MP}}{T_i^{SP}}, \tag{2.5}$$

ANTT is a lower-is-better metric. For the above example, the one program achieves an NTT of $1/0.75 = 1.33$ and the other $1/0.5 = 2$, and thus ANTT equals $(1.33 + 2)/2 = 1.67$, which means that the average slowdown per program equals 1.67.

2.3.3 COMPARISON TO PREVALENT METRICS

Prevalent metrics for quantifying multiprogram performance in the architecture community are IPC throughput, weighted speedup [174] and harmonic mean [129]. We will now discuss these performance metrics and compare them against STP and ANTT.

Computer architects frequently use single-threaded SPEC CPU benchmarks to compose multiprogram workloads for which IPC is an adequate performance metric. Hence, the multiprogram metrics that have been proposed over the recent years are based on IPC (or CPI). This limits the applicability of these metrics to single-threaded workloads. STP and ANTT, on the other hand, are defined in terms of execution time, which makes the metrics applicable for multi-threaded workloads as well. In order to be able to compare STP and ANTT against the prevalent metrics, we first convert STP and ANTT to CPI-based metrics based on Equation 2.1 while assuming that the number of instructions and clock frequency is constant. It easily follows that STP can be computed as

$$STP = \sum_{i=1}^{n} \frac{CPI_i^{SP}}{CPI_i^{MP}}, \tag{2.6}$$

with CPI_i^{SP} and CPI_i^{MP} the CPI under single-program and multiprogram execution, respectively. ANTT can be computed as

$$ANTT = \frac{1}{n} \sum_{i=1}^{n} \frac{CPI_i^{MP}}{CPI_i^{SP}}. \tag{2.7}$$

IPC throughput.   IPC throughput is defined as the sum of the IPCs of the co-executing programs:

$$IPC\_throughput = \sum_{i=1}^{n} IPC_i. \tag{2.8}$$

IPC throughput naively reflects a computer architect's view on throughput; however, it doesn't have a meaning in terms of performance from either a user perspective or a system perspective. In particular, one could optimize a system's IPC throughput by favoring high-IPC programs; however, this may not necessarily reflect improvements in system-level performance (job turnaround time and/or system throughput). Therefore, it should not be used as a multiprogram performance metric.

Weighted speedup.   Snavely and Tullsen [174] propose weighted speedup to evaluate how well jobs co-execute on a multi-threaded processor. Weighted speedup is defined as

$$weighted\_speedup = \sum_{i=1}^{n} \frac{IPC_i^{MP}}{IPC_i^{SP}}, \tag{2.9}$$

The motivation by Snavely and Tullsen for using IPC as the basis for the speedup metric is that if one job schedule executes more instructions than another in the same time interval, it is more symbiotic and, therefore, yields better performance; the weighted speedup metric then equalizes the contribution of each program in the job mix by normalizing its multiprogram IPC with its single-program IPC.

From the above, it follows that weighted speedup equals system throughput (STP) and, in fact, has a physical meaning — it relates to the number of jobs completed per unit of time — although this may not be immediately obvious from weighted speedup's definition and its original motivation.

Harmonic mean.   Luo et al. [129] propose the harmonic mean metric, or hmean for short, which computes the harmonic mean rather than an arithmetic mean (as done by weighted speedup) across the IPC speedup numbers:

$$hmean = \frac{n}{\sum_{i=1}^{n} \frac{IPC_i^{SP}}{IPC_i^{MP}}}. \tag{2.10}$$

The motivation by Luo et al. for computing the harmonic mean is that it tends to result in lower values than the arithmetic average if one or more programs have a lower IPC speedup, which they argue better captures the notion of fairness than weighted speedup. The motivation is based solely on properties of the harmonic and arithmetic means and does not reflect any system-level meaning. It follows from the above that the hmean metric is the reciprocal of the ANTT metric, and hence it has a system-level meaning, namely, it relates to (the reciprocal of) the average job's normalized turnaround time.

2.3.4   STP VERSUS ANTT PERFORMANCE EVALUATION

Because of the complementary perspectives, multiprogram workload performance should be characterized using both the STP and ANTT metrics in order to get a more comprehensive performance picture. Multiprogram performance, using only one metric provides an incomplete view and skews the perspective. Eyerman and Eeckhout [61] illustrate this point by comparing different SMT fetch policies: one fetch policy may outperform another fetch policy according to one metric; however, according to the other metric, the opposite may be true. Such a case illustrates that there is trade-off in user-level performance versus system-level performance. This means that one fetch policy may yield higher system throughput while sacrificing average job turnaround time, whereas the other one yields shorter average job turnaround times while reducing system throughput. So, it is important to report both the STP and ANTT metrics when reporting multiprogram performance.

2.4   AVERAGE PERFORMANCE

So far, we discussed performance metrics for individual workloads only. However, people like to quantify what this means for average performance across a set of benchmarks. Although everyone agrees that 'Performance is not a single number', there has been (and still is) a debate going on about which number is better. Some argue for the arithmetic mean, others argue for the harmonic mean, yet others argue for the geometric mean. This debate has a long tradition: it started in 1986 with Fleming and Wallace [68] arguing for the geometric mean. Smith [173] advocated the opposite, shortly thereafter. Cragon [37] also argues in favor of the arithmetic and harmonic mean. Hennessy and Patterson [80] describe the pros and cons of all three averages. More recently, John [93] argued strongly against the geometric mean, which was counterattacked by Mashey [134].

It seems that even today people are still confused about which average to choose; some papers use the harmonic mean, others use the arithmetic mean, and yet others use the geometric mean. The goal of this section is to shed some light into this important problem, describe the two opposite viewpoints and make a recommendation. The contradictory views come from approaching the problem from either a mathematical angle or a statistical angle. The mathematical viewpoint, which leads to using the harmonic and arithmetic mean, starts from understanding the physical meaning of the performance metric, and then derives the average in a meaningful way that makes sense mathematically. The statistical viewpoint, on the other hand, assumes that the benchmarks are randomly chosen from the workload population, and the performance speedup metric is log-normally distributed, which leads to using the geometric mean.

2.4.1   HARMONIC AND ARITHMETIC AVERAGE: MATHEMATICAL VIEWPOINT

The mathematical viewpoint starts from a clear understanding of what the metrics mean, and then chooses the appropriate average (arithmetic or harmonic) for the metrics one wants to compute the average for. This approach does not assume a particular distribution for the underlying population, and it does not assume that the benchmarks are chosen randomly from the workload space. It simply computes the average performance metric for the selected set of benchmarks in a way that makes sense physically, i.e., understanding the physical meaning of the metric one needs to compute the average for, leads to which average to use (arithmetic or harmonic) in a mathematically meaningful way.

In particular, if the metric of interest is obtained by dividing A by B and if A is weighed equally among the benchmarks then the harmonic mean is meaningful. Here is the mathematical derivation:

$$\frac{\sum_{i=1}^{n} A_i}{\sum_{i=1}^{n} B_i} = \frac{n \cdot A}{\sum_{i=1}^{n} B_i} = \frac{n}{\sum_{i=1}^{n} \frac{B_i}{A}} = \frac{n}{\sum_{i=1}^{n} \frac{1}{A/B_i}} = \frac{n}{\sum_{i=1}^{n} \frac{1}{A_i/B_i}} = HM(A_i/B_i). \qquad (2.11)$$

If on the other hand, B is weighed equally among the benchmarks, then the arithmetic mean is meaningful:

$$\frac{\sum_{i=1}^{n} A_i}{\sum_{i=1}^{n} B_i} = \frac{\sum_{i=1}^{n} A_i}{n \cdot B} = \frac{1}{n} \sum_{i=1}^{n} \frac{A_i}{B} = \frac{1}{n} \sum_{i=1}^{n} \frac{A_i}{B_i} = AM(A_i/B_i). \qquad (2.12)$$

We refer to John [93] for a more extensive description, including a discussion on how to weigh the different benchmarks.

Hence, depending on the performance metric and how the metric was computed, one has to choose for either the harmonic or arithmetic mean. For example, assume we have selected a 100M instruction sample for each benchmark in our benchmark suite. The average IPC (instructions executed per cycles) needs to be computed as the harmonic mean across the IPC numbers for the individual benchmarks because the instruction count is constant across the benchmarks. The same applies to computing MIPS or million instructions executed per second. Inversely, the average CPI (cycles per instruction) needs to be computed as the arithmetic mean across the individual CPI numbers. Similarly, the arithmetic average also applies for TPI (time per instruction).

The choice for harmonic versus arithmetic mean also depends on the experimenter's perspective. For example, when computing average speedup (execution time on original system divided by execution time on improved system) over a set of benchmarks, one needs to use the harmonic mean if the relative duration of the benchmarks is irrelevant, or, more precisely, if the experimenter weighs the time spent in the original system for each of the benchmarks equally. If on the other hand, the experimenter weighs the duration for the individual benchmarks on the enhanced system equally, or if one expects a workload in which each program will run for an equal amount of time on the enhanced system, then the arithmetic mean is appropriate.

Weighted harmonic mean and weighted arithmetic can be used if one knows a priori which applications will be run on the target system and in what percentages — this may be the case in some embedded systems. Assigning weights to the applications proportional to their percentage of execution will provide an accurate performance assessment.

2.4.2   GEOMETRIC AVERAGE: STATISTICAL VIEWPOINT

The statistical viewpoint regards the set of benchmarks as being representative for a broader set of workloads and assumes that the population from which the sample (the benchmarks) are taken follows some distribution. In particular, Mashey [134] argues that speedups (the execution time on the reference system divided by the execution time on the enhanced system) are distributed following a log-normal distribution. A log-normal distribution means that the elements in the population are not normally or Gaussian distributed, but their logarithms (of any base) are. The main difference between a normal and a log-normal distribution is that a log-normal distribution is skewed, i.e., its distribution is asymmetric and has a long tail to the right, whereas a normal distribution is symmetric and has zero skew. Normal distributions arise from aggregations of many additive effects, whereas log-normal distributions arise from combinations of multiplicative effects. In fact, performance is a multiplicative effect in CPI and clock frequency for a given workload, see Equation 2.1, which may seem to suggest that a log-normal distribution is a reasonable assumption; however, CPI and clock frequency are not independent, i.e., the micro-architecture affects both CPI and clock frequency. The assumption that speedups are distributed along a log-normal distribution also implies that some programs experience much larger speedups than others, i.e., there are outliers, hence the long tail to the right.

Assuming that speedups follow a log-normal distribution, the geometric mean is the appropriate mean for speedups:

$$ GM = \sqrt[1/n]{\prod_{i=1}^{n} x_i} = \exp\left(\frac{1}{n}\sum_{i=1}^{n}\ln(x_i)\right). \tag{2.13} $$

In this formula, it is assumed that x is log-normally distributed, or, in other words, $\ln(x)$ is normally distributed. The average of a normal distribution is the arithmetic mean; hence, the exponential of the arithmetic mean over $\ln(x_i)$ computes the average for x (see the right-hand side of the above formula). This equals the definition of the geometric mean (see the left-hand side of the formula).

The geometric mean has an appealing property. One can compute the average speedup between two machines by dividing the average speedups for these two machines relative to some reference machine. In particular, having the average speedup numbers of machines A and B relative to some reference machine R, one can compute the average relative speedup between A and B by simply dividing the former speedup numbers. SPEC CPU uses the geometric mean for computing average SPEC rates, and the speedups are computed against some reference machine, namely a Sun Ultra5_10 workstation with a 300MHz SPARC processor and 256MB main memory.

The geometric mean builds on two assumptions. For one, it assumes that the benchmarks are representative for the much broader workload space. A representative set of benchmarks can be obtained by randomly choosing benchmarks from the population, provided a sufficiently large number of benchmarks are taken. Unfortunately, the set of benchmarks is typically not chosen randomly from a well-defined workload space; instead, a benchmark suite is a collection of interesting benchmarks covering an application domain of interest picked by a committee, an individual or a marketing organization. In other words, and as argued in the introduction, picking a set of workloads is subject to the experimenter's judgment and experience. Second, the geometric mean assumes that the speedups are distributed following a log-normal distribution. These assumptions have never been validated (and it is not clear how they can ever be validated), so it is unclear whether these assumptions holds true. Hence, given the high degree of uncertainty regarding the required assumptions, using the geometric mean for computing average performance numbers across a set of benchmarks is of questionable value.

2.4.3 FINAL THOUGHT ON AVERAGES

The above reasoning about how to compute averages reveals a basic problem. Harmonic mean (and arithmetic mean) can provide a meaningful summary for a given set of benchmarks, but then extrapolating that performance number to a full workload space is questionable. Geometric mean suggests a way of extrapolating performance to a full workload space, but the necessary assumptions are unproven (and are probably not valid). The one important exception is when the full workload space is known – as would be the case for some embedded systems. In this case, a weighted harmonic mean (or weighted arithmetic mean) would work well.

2.5 PARTIAL METRICS

The metrics discussed so far consider overall performance, i.e., the metrics are based on total execution time. Partial metrics, such as cache miss ratio, branch misprediction rate, misses per thousand instructions, or bus utilization, reveal insight in what sources affect processor performance. Partial metrics allow for studying particular components of a processor in isolation.

CHAPTER 3

Workload Design

Workload design is an important step in computer architecture research, as already described in the introduction: subsequent steps in the design process are subject to the selection process of benchmarks, and choosing a non-representative set of benchmarks may lead to biased observations, incorrect conclusions, and, eventually, designs that poorly match their target workloads.

For example, Maynard et al. [137] as well as Keeton et al. [105] compare the behavior of commercial applications, including database servers, against the SPEC CPU benchmarks that are widely used in computer architecture. They find that commercial workloads typically exhibit more complex branch behavior, larger code and data footprints, and more OS as well as I/O activity. In particular, the instruction cache footprint of the SPEC CPU benchmarks is small compared to commercial workloads; also, memory access patterns for footprints that do not fit in on-chip caches are typically regular or strided. Hence, SPEC CPU benchmarks are well suited for pipeline studies, but they should be used with care for memory performance studies. Guiding processor design decisions based on the SPEC CPU benchmarks only may lead to suboptimal performance for commercial workloads.

3.1 FROM WORKLOAD SPACE TO REPRESENTATIVE WORKLOAD

Ideally, we would like to have a small set of benchmarks that is representative for the broader workload space, see also Figure 3.1. The *workload space* is defined as the collection of all possible applications. Although it may be possible to enumerate all the applications that will run on a given device in some application domains, i.e., in some embedded systems, this is not possible in general, e.g., the general-purpose domain. The *reference workload* then is the set of benchmarks that the experimenter believes to be representative for the broader workload space. In the embedded system example, it may be possible to map applications to benchmarks one-to-one and have a reference workload that very well represents the workload space. In the general-purpose computing domain, constructing the reference workload is subject to the experimenter's judgment and experience. Because the number of benchmarks in the reference workload can be large, it may be appropriate to reduce the number of benchmarks towards a *reduced workload*. The bulk of this chapter deals with workload reduction methodologies which reduce the number of benchmarks in the reference workload to a limited number of benchmarks in the reduced workload. However, before doing so, we first discuss the difficulty in coming up with a good reference workload.

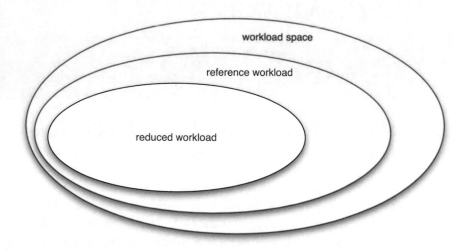

Figure 3.1: Introducing terminology: workload space, reference workload and reduced workload.

3.1.1   REFERENCE WORKLOAD

Coming up with a good reference workload is a difficult problem for at least three reasons. For one, the workload space may be huge (and is typically not even well-defined). There exist many different, important, application domains, such as general-purpose computing, media (e.g., audio and video codecs), scientific computing, bioinformatics, medical, commercial (e.g., database, web servers, mail servers, e-commerce). And each of these domains has numerous applications and thus numerous potential benchmarks that can be derived from these applications. As a result, researchers have come up with different benchmark suites to cover different application domains, e.g., SPEC CPU (general-purpose), MediaBench [124] (media), and BioPerf (bioinformatics) [8]. In addition, new benchmark suites emerge to evaluate new technologies, e.g., PARSEC [15] to evaluate multi-threaded workloads on multicore processors, DaCapo [18] to evaluate Java workloads, or STAMP [141] to evaluate transactional memory. In addition to this large number of existing application domains and benchmark suites, new application domains keep on emerging. For example, it is likely to expect that the trend towards cloud computing will change the workloads that will run on future computer systems.

Second, the workload space is constantly changing, and, hence, the design may be optimized for a workload that is irrelevant (or at least, less important) by the time the product hits the market. This is a constant concern computer architects need to consider: given that it is unknown what the future's workloads will be, architects are forced to evaluate future systems using existing benchmarks, which are often modeled after yesterday's applications. In addition, architects also have to rely on old compiler technology for generating the binaries. To make things even worse, future instruction-set architectures may be different than ones available today, which requires re-compilation and which may lead to different performance numbers; also, the workloads (e.g., transaction processing workloads such as databases) may need to be re-tuned for the system under study. In summary, one could say that architects are designing tomorrow's systems using yesterday's benchmarks and tools. To evaluate the impact of workload and tool drift, Yi et al. [198] describe an experiment in which they optimize a superscalar processor using SPEC CPU95 benchmarks and then evaluate the performance and energy-efficiency using SPEC CPU2000. They conclude that the CPU95 optimized processor performs 20% worse compared to the CPU2000 optimized design, the primary reason being that CPU2000 is more memory-intensive than CPU95. In conclusion, architects should be well aware of the impact workload drift may have, and they should thus anticipate future workload behavior as much as possible in order to have close to optimal performance and efficiency for future workloads running on future designs.

Finally, the process of selecting benchmarks to be included in a benchmark suite itself is subjective. John Henning [81] describes the selection process used by SPEC when composing the CPU2000 benchmark suite. The Standard Performance Evaluation Corporation (SPEC) is a nonprofit consortium whose members include hardware vendors, software vendors, universities, customers, and consultants. The SPEC subcommittee in charge of the CPU2000 benchmarks based its decisions on a number of criteria, ranging from portability and maintainability of the benchmarks' source code, to performance characteristics of the benchmarks, to vendor interest. Hence, it is hard to argue that this selection process reflects an objective draw of a scientific sample from a population.

3.1.2   TOWARDS A REDUCED WORKLOAD

The large number of application domains and the large number of potential benchmarks per application domain leads to a huge number of potential benchmarks in the reference workload. Simulating all of these benchmarks is extremely time-consuming and is often impractical or infeasible. Hence, researchers have studied methods for reducing the number of benchmarks in order to save simulation time. Citron [33] performed a wide survey on current practice in benchmark subsetting. He found that selecting a limited number of benchmarks from a benchmark suite based on programming language, portability and simulation infrastructure concerns is common and may lead to misleading performance numbers. Hence, picking a reduced set of benchmarks should not be done in an ad-hoc way.

This chapter describes two methodologies for composing a reduced but representative benchmark suite from a larger set of specified benchmarks, the reference workload. In other words, these methodologies seek to reduce the amount of work that needs to be done during performance studies while retaining meaningful results. The basic idea is to compose a reduced workload by picking a limited but representative set of benchmarks from the reference workload. The motivation for reducing the number of benchmarks in the workload is to limit the amount of work that needs to be done during performance evaluation. Having too many benchmarks only gets in the way: redundant benchmarks, i.e., benchmarks that exhibit similar behaviors and thus stress similar aspects of the design as other benchmarks, only increase experimentation time without providing any additional insight.

The two methodologies described in this chapter are based on statistical data analysis techniques. The first approach uses Principal Component Analysis (PCA) for understanding the (dis)similarities across benchmarks, whereas the second approach uses the Plackett and Burman design of experiment. The fundamental insight behind both methodologies is that different benchmarks may exhibit similar behaviors and stress similar aspects of the design. As a result, there is no need to include such redundant benchmarks in the workload.

3.2   PCA-BASED WORKLOAD DESIGN

The first workload reduction methodology that we discuss is based on Principal Component Analysis (PCA) [96]. PCA is a well-known statistical data analysis technique that reduces the dimensionality of a data set. More precisely, it transforms a number of possibly correlated variables (or dimensions) into a smaller number of uncorrelated variables, which are called the principal components. Intuitively speaking, PCA has the ability to describe a huge data set along a limited number of dimensions. In other words, it presents the user with a lower-dimensional picture that yet captures the essence of the more-dimensional data set.

Eeckhout et al. [53] describe how to leverage this powerful technique for analyzing workload behavior. They view the workload space as a p-dimensional space with p the number of important program characteristics. These program characteristics describe various behavioral aspects of a workload, such as instruction mix, amount of instruction-level parallelism (ILP), branch predictability, code footprint, memory working set size, and memory access patterns. Because the software and hardware are becoming more complex, the number of program characteristics p is large in order to capture a meaningful behavioral characterization. As a result, p is typically too large to easily visualize and/or reason about the workload space — visualizing a 20-dimensional space is not trivial. In addition, correlation may exist between these program characteristics. For example, complex program behavior may reflect in various metrics such as poor branch predictability, large code and memory footprint, and complex memory access patterns. As a result, a high value for one program characteristic may also imply a high value for another program characteristic. Correlation along program characteristics complicates the understanding: one may think two workloads are different from each other because they seem to be different along multiple dimensions; however, the difference may be due to a single underlying mechanism that reflects itself in several (correlated) program characteristics.

The large number of dimensions in the workload space and the correlation among the dimensions complicates the understanding substantially. PCA transforms the p-dimensional workload space to a q-dimensional space (with $q \ll p$) in which the dimensions are uncorrelated. The transformed space thus provides an excellent opportunity to understand benchmark (dis)similarity. Benchmarks that are far away from each other in the transformed space show dissimilar behavior; benchmarks that are close to each other show similar behavior. When combined with cluster analysis (CA) as a subsequent step to pick to most diverse benchmarks in the reference workload, PCA can determine a reduced but representative workload.

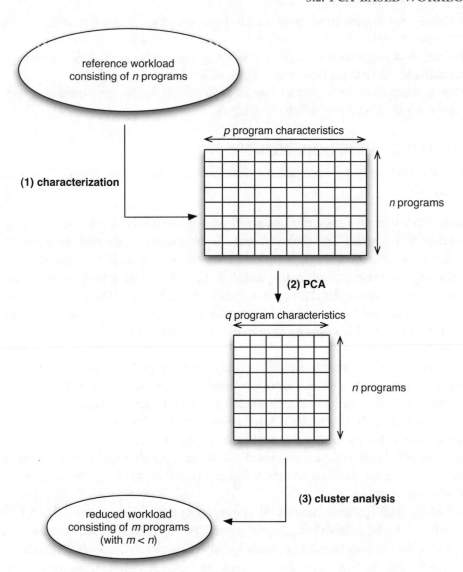

Figure 3.2: Schematic overview of the PCA-based workload reduction method.

3.2.1   GENERAL FRAMEWORK

Figure 3.2 illustrates the general framework of the PCA-based workload reduction method. It starts off from a set of benchmarks, called the reference workload. For each of these benchmarks, it then collects a number of program characteristics. This yields a large data matrix with as many rows as there are benchmarks, namely n, and with as many columns as there are program characteristics, namely p. Principal component analysis then transforms the p program characteristics into q principal components, yielding an $n \times q$ data matrix. Subsequently, applying cluster analysis identifies m representative benchmarks out of the n benchmarks (with $m < n$).

The next few sections describe the three major steps in this methodology: workload characterization, principal component analysis and cluster analysis.

3.2.2   WORKLOAD CHARACTERIZATION

The first step is to characterize the behavior of the various benchmarks in the workload. There are many ways for doing so.

Hardware performance monitors.   One way of characterizing workload behavior is to employ hardware performance monitors — in fact, hardware performance monitors are widely used (if not prevailing) in workload characterization because they are available on virtually all modern microprocessors, can measure a wide range of events, are easy to use, and allow for characterizing long-running complex workloads that are not easily simulated (i.e., the overhead is virtually zero). The events measured using hardware performance monitors are typically instruction mix (e.g., percentage loads, branches, floating-point operations, etc.), IPC (number of instructions retired per cycle), cache miss rates, branch mispredict rates, etc.

Inspite of its widespread use, there is a pitfall in using hardware performance monitors: they can be misleading in the sense that they can conceal the workload's inherent behavior. This is to say that different inherent workload behavior can lead to similar behavior when measured using hardware performance monitors. As a result, based on a characterization study using hardware performance monitors one may conclude that different benchmarks exhibit similar behavior because they show similar cache miss rates, IPC, branch mispredict rates, etc.; however, a more detailed analysis based on a microarchitecture-independent characterization (as described next) shows that both benchmarks exhibit different inherent behavior.

Hoste and Eeckhout [85] present data that illustrates exactly this pitfall, see also Table 3.1 for an excerpt. The two benchmarks, gzip with the graphic input from the SPEC CPU2000 benchmark suite and fasta from the BioPerf benchmark suite, exhibit similar behavior in terms of CPI and cache miss rates, as measured using hardware performance monitors. However, the working set size and memory access patterns are shown to be very different. The data working set size is an order of magnitude bigger for gzip compared to fasta, and the memory access patterns seem to be fairly different as well between these workloads.

The notion of microarchitecture-independent versus microarchitecture-dependent characterization also appears in sampled simulation, see Chapter 6.

Hardware performance monitor data across multiple machines.   One way for alleviating this pitfall is to characterize the workload on a multitude of machines and architectures. Rather than collecting hardware performance monitor data on a single machine, collecting data across many different

Table 3.1: This case study illustrates the pitfall in using microarchitecture-dependent performance characteristics during workload characterization: although microarchitecture-dependent characteristics may suggest that two workloads exhibit similar behavior, this may not be the case when looking into the microarchitecture-independent characteristics.

Microarchitecture-dependent characterization		
	gzip-graphic	fasta
CPI on Alpha 21164	1.01	0.92
CPI on Alpha 21264	0.63	0.66
L1 D-cache misses per instruction	1.61%	1.90%
L1 I-cache misses per instruction	0.15%	0.18%
L2 cache misses per instruction	0.78%	0.25%
Microarchitecture-independent characterization		
	gzip-graphic	fasta
Data working set size (# 4KB pages)	46,199	4,058
Instruction working set size (# 4KB pages)	33	79
Probability for two consecutive dynamic executions of the same static load to access the same data	0.67	0.30
Probability for two consecutive dynamic executions of the same static store to access the same data	0.64	0.05
Probability for the difference in memory addresses between two consecutive loads in the dynamic instruction stream to be smaller than 64	0.26	0.18
Probability for the difference in memory addresses between two consecutive stores in the dynamic instruction stream to be smaller than 64	0.35	0.93

machines is likely to yield a more comprehensive and more informative workload characterization because different machines and architectures are likely to stress the workload behavior (slightly) differently. As a result, an inherent behavioral difference between benchmarks is likely to show up on at least one of a few different machines.

Phansalkar et al. [160] describe an experiment in which they characterize the SPEC CPU2006 benchmark suite on five different machines with four different ISAs and compilers (IBM Power, Sun UltraSPARC, Itanium and x86). They use this multi-machine characterization as input for the PCA-based workload analysis method, and then study the diversity among the benchmarks in the SPEC CPU2006 benchmark suite. This approach was used by SPEC for the development of the CPU2006 benchmark suite [162]: the multi-machine workload characterization approach was used to understand the diversity and similarity among the benchmarks for potential inclusion in the CPU2006 benchmark suite.

Table 3.2: Example microarchitecture-independent characteristics that can be used as input for the PCA-based workload reduction method.

Program Characteristic	Description
instruction mix	Percentage of loads, stores, branches, integer arithmetic operations, floating-point operations.
instruction-level parallelism (ILP)	Amount of ILP for a given window of instructions, e.g., the IPC achieved for an idealized out-of-order processor (with perfect branch predictor and caches).
data working set size	The number of unique memory blocks or pages touched by the data stream.
code footprint	The number of unique memory blocks or pages touched by the instruction stream.
branch transition rate	The number of times that a branch switches between taken and not-taken directions during program execution.
data stream strides	Distribution of the strides observed between consecutive loads or stores in the dynamic instruction stream. The loads or stores could be consecutive executions of the same static instruction (local stride) or could be consecutive instructions from whatever load or store (global stride).

Detailed simulation.   One could also rely on detailed cycle-accurate simulation for collecting program characteristics in a way similar to using hardware performance monitors. The main disadvantage is that is extremely time-consuming to simulate industry-standard benchmarks in a cycle-accurate manner — cycle-accurate simulation is at least five orders of magnitude slower than native hardware execution. The benefit though is that simulation enables collecting characteristics on a range of machine configurations that are possibly not (yet) available.

Microarchitecture-independent workload characterization.   Another way for alleviating the hardware performance monitor pitfall is to collect a number of program characteristics that are independent of a specific microarchitecture. The key benefit of a microarchitecture-independent characterization is that it is not biased towards a specific hardware implementation. Instead, it characterizes a workload's 'inherent' behavior (but is still dependent on the instruction-set architecture and compiler). The disadvantage of microarchitecture-independent characterization is that most of the characteristics can only be measured through software. Although measuring these characteristics is done fairly easily through simulation or through binary instrumentation (e.g., using a tool like Atom [176] or Pin [128]), it can be time-consuming because the simulation and instrumentation may incur a slowdown of a few orders of magnitude — however, it will be much faster than detailed cycle-accurate simulation. In practice though, this may be a relatively small concern because the workload characterization is a one-time cost.

Example microarchitecture-independent characteristics are shown in Table 3.2. They include characteristics for code footprint, working set size, branch transition rate, and memory access patterns (both local and global). See the work by Joshi et al. [99] and Hoste and Eeckhout [85] for more detailed examples.

3.2.3   PRINCIPAL COMPONENT ANALYSIS

The second step in the workload reduction method is to apply principal components analysis (PCA) [96]. PCA is a statistical data analysis technique that transforms a number of possibly correlated variables (i.e., program characteristics) in a smaller number of uncorrelated *principal components*.

The principal components are linear combinations of the original variables, and they are uncorrelated. Mathematically speaking, PCA transforms p variables $X_1, X_2, \ldots, X_p$ into p principal components $Z_1, Z_2, \ldots, Z_p$ with $Z_i = \sum_{j=1}^{p} a_{ij} X_j$. This transformation is done such that the first principal component shows the greatest variance, followed by the second, and so forth: $Var[Z_1] > Var[Z_2] > \ldots > Var[Z_p]$. Intuitively speaking, this means that the first principal component Z_1 captures the most 'information' and the final principal component Z_p the least. In addition, the principal components are uncorrelated, or $Cov[Z_i, Z_j] = 0, \forall i \neq j$, which basically means that there is no information overlap between the principal components. Note that the total variance in the data remains the same before and after the transformation, namely $\sum_{i=1}^{p} Var[X_i] = \sum_{i=1}^{p} Var[Z_i]$. Computing the principal components is done by computing the eigenvalue decomposition of a data covariance matrix. Figure 3.3 illustrates how PCA operates in a two-dimensional space on a Gaussian distributed data set. The first principal component is determined by the direction with the maximum variance; the second principal component is orthogonal to the first one.

Because the first few principal components capture most of the information in the original data set, one can drop the trailing principal components with minimal loss of information. Typically, the number of retained principal components q is much smaller than the number of dimensions p in the original data set. The amount of variance $(\sum_{i=1}^{q} Var[Z_i])/(\sum_{i=1}^{p} Var[X_i])$ accounted for by the retained principal components, provides a measure for the amount of information retained after PCA. Typically, over 80% of the total variance is explained by the retained principal components.

PCA can be performed by most existing statistical software packages, both in commercial packages such as SPSS, SAS and S-PLUS, as well as open-source packages such as R.

3.2.4   CLUSTER ANALYSIS

The end result from PCA is a data matrix with n rows (the benchmarks) and q columns (the principal components). Cluster analysis now groups (or clusters) the n benchmarks based on the q principal components. The final goal is to obtain a number of clusters, with each cluster grouping a set of benchmarks that exhibit similar behavior. There exist two common clustering techniques, namely agglomerative hierarchical clustering and K-means clustering.

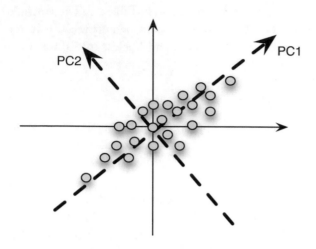

Figure 3.3: PCA identifies the principal components in a data set.

Agglomerative hierarchical clustering considers each benchmark as a cluster initially. In each iteration of the algorithm, the two clusters that are closest to each other are grouped to form a new cluster. The distance between the merged clusters is called the *linkage distance*. Nearby clusters are progressively merged until finally all benchmarks reside in a single big cluster. This clustering process can be represented in a so called *dendrogram*, which graphically represents the linkage distance for each cluster merge. Having obtained a dendrogram, it is up to the user to decide how many clusters to retain based on the linkage distance. Small linkage distances imply similar behavior in the clusters, whereas large linkage distances suggest dissimilar behavior. There exist a number of methods for calculating the distance between clusters — the inter-cluster distance is needed in order to know which clusters to merge. For example, the furthest neighbor method (also called complete-linkage clustering) computes the largest distance between any two benchmarks in the respective clusters; in average-linkage clustering, the inter-cluster distance is computed as the average distance.

K-means clustering starts by randomly choosing k cluster centers. Each benchmark is then assigned to its nearest cluster, and new cluster centers are computed. The next iteration then re-computes the cluster assignments and cluster centers. This iterative process is repeated until some convergence criterion is met. The key advantage of K-means clustering is its speed compared to hierarchical clustering, but it may lead to different clustering results for different initial random assignments.

3.2.5 APPLICATIONS

The PCA-based methodology enables various applications in workload characterization.

Workload analysis.   Given the limited number of principal components, one can visualize the workload space, as illustrated in Figure 3.4, which shows the PCA space for a set of SPEC CPU95 benchmarks along with TPC-D running on the postgres DBMS. (This graph is based on the data presented by Eeckhout et al. [53] and represents old and obsolete data — both CPU95 and TPC-D are obsolete — nevertheless, it illustrates various aspects of the methodology.) The graphs show the various benchmarks as dots in the space spanned by the first and second principal components, and third and fourth principal components, respectively. Collectively, these principal components capture close to 90% of the total variance, and thus they provide an accurate picture of the workload space. The different colors denote different benchmarks; the different dots per benchmark denote different inputs.

By interpreting the principal components, one can reason about how benchmarks differ from each other in terms of their execution behavior. The first principal component primarily quantifies a benchmark's control flow behavior: benchmarks with relatively few dynamically executed branch instructions and relatively low I-cache miss rates show up with a high first principal component. One example benchmark with a high first principal component is ijpeg. Benchmarks with high levels of ILP and poor branch predictability have a high second principal component, see for example go and compress. The third and fourth primarily capture D-cache behavior and the instruction mix, respectively.

Several interesting observations can be made from these plots. Some benchmarks exhibit execution behavior that is different from the other benchmarks in the workload. For example, ijpeg, go and compress seem to be isolated in the workload space and seem to be relatively dissimilar from the other benchmarks. Also, the inputs given to the benchmark may have a big impact for some benchmarks, e.g., TPC-D, whereas for other benchmarks the execution behavior is barely affected by its input, e.g., ijpeg. The different inputs (queries) are scattered around for TPC-D; hence, different inputs seem to lead to fairly dissimilar behavior; for ijpeg, on the other hand, the inputs seem to clustered, and inputs seem to have limited effect on the program's execution behavior. Finally, it also suggests that this set of benchmarks only partially covers the workload space. In particular, a significant part of the workload space does not seem to be covered by the set of benchmarks, as illustrated in Figure 3.5.

Workload reduction.   By applying cluster analysis after PCA, one can group the various benchmarks into a limited number of clusters based on their behavior, i.e., benchmarks with similar execution behavior are grouped in the same cluster. The benchmark closest to the cluster's centroid can then serve as the representative for the cluster. By doing so, one can reduce the workload to a limited set of representative benchmarks — also referred to as benchmark subsetting. Phansalkar et al. [160] present results for SPEC CPU2006, and they report average prediction errors of 3.8% and 7% for the subset compared to the full integer and floating-point benchmark suite, respectively, across five commercial processors. Table 3.3 summarizes the subsets for the integer and floating-point benchmarks.

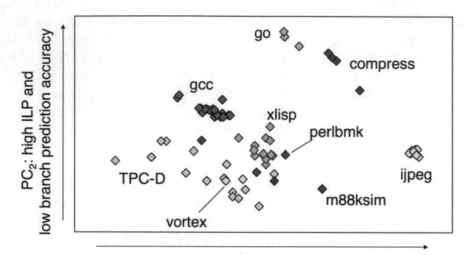

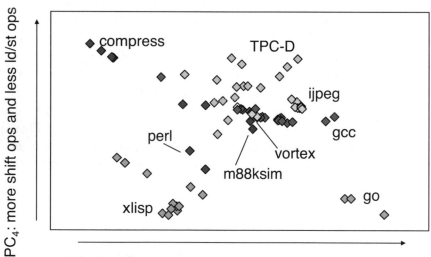

Figure 3.4: Example PCA space as a function of the first four principal components: the first and second principal components are shown in the top graph, and the third and fourth principal components are shown in the bottom graph.

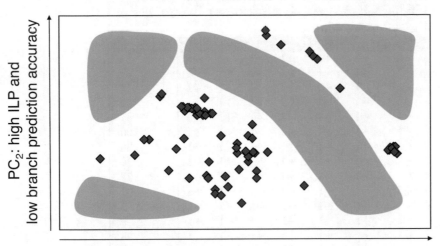

Figure 3.5: The PCA-based workload analysis method allows for finding regions (weak spots) in the workload space that are not covered by a benchmark suite. Weak spots are shaded in the graph.

Table 3.3: Representative SPEC CPU2006 subsets according to the study done by Phansalkar et al. [160].	
SPEC CINT2006	400.perlbench, 471.omnetpp, 429.mcf, 462.libquantum, 473.astar, 483.xalancbmk
SPEC CFP2006	437.leslie3d, 454.calculix, 459.GemsFDTD, 436.cactusADM, 447.dealII, 450.soplex, 470.lbm, 453.povray

Other applications.   The PCA-based methodology has been used for various other or related purposes, including evaluating the DaCapo benchmark suite [18], analyzing workload behavior over time [51], studying the interaction between the Java application, its input and the Java virtual machine (JVM) [48], and evaluating the representativeness of (reduced) inputs [52].

3.3   PLACKETT AND BURMAN BASED WORKLOAD DESIGN

Yi et al. [196] describe a simulation-friendly approach for understanding how workload performance is affected by microarchitecture parameters. They therefore employ a Plackett and Burman (PB) design of experiment which involves a small number of simulations — substantially fewer simulations than simulating all possible combinations of microarchitecture parameters — while still capturing the effects of each parameter and selected interactions. Also, it provides more information compared to a one-at-a-time experiment for about the same number of simulations: PB captures interaction effects which a one-at-a-time experiment does not. In particular, a Plackett and Burman design

Table 3.4: An example Plackett and Burman design matrix with foldover.

A	B	C	D	E	F	G	exec. time
+1	+1	+1	−1	+1	−1	−1	56
−1	+1	+1	+1	−1	+1	−1	69
−1	−1	+1	+1	+1	−1	+1	15
+1	−1	−1	+1	+1	+1	−1	38
−1	+1	−1	−1	+1	+1	+1	45
+1	−1	+1	−1	−1	+1	+1	100
+1	+1	−1	+1	−1	−1	+1	36
−1	−1	−1	−1	−1	−1	−1	20
−1	−1	−1	+1	−1	+1	+1	77
+1	−1	−1	−1	+1	−1	+1	87
+1	+1	−1	−1	−1	+1	−1	5
−1	+1	+1	−1	−1	−1	+1	9
+1	−1	+1	+1	−1	−1	−1	28
−1	+1	−1	+1	+1	−1	−1	81
−1	−1	+1	−1	+1	+1	−1	67
+1	+1	+1	+1	+1	+1	+1	2
−31	−129	−44	−44	47	71	7	

(with foldover) involves $2c$ simulations to quantify the effect of c microarchitecture parameters and all pairwise interactions. The outcome of the PB experiment is a ranking of the most significant microarchitecture performance bottlenecks. Although the primary motivation for Yi et al. to propose the Plackett and Burman design was to explore the microarchitecture design space in a simulation-friendly manner, it also has important applications in workload characterization. The ranking of performance bottlenecks provides a unique signature that characterizes a benchmark in terms of how it stresses the microarchitecture. By comparing bottleneck rankings across benchmarks one can derive how (dis)similar the benchmarks are.

The Plackett and Burman design uses a design matrix — Table 3.4 shows an example design matrix with foldover — Yi et al. and the original paper by Plackett and Burman provide design matrices of various dimensions. A row in the design matrix corresponds to a microarchitecture configuration that needs to be simulated; each column denotes a different microarchitecture parameter. A '+1' and '−1' value represents a high and low — or on and off — value for a parameter. For example, a high and low value could be a processor width of 8 and 2, respectively; or with aggressive hardware prefetching and without prefetching, respectively. It is advised that the high and low values be just outside of the normal or expected range of values in order to take into account the full potential impact of a parameter. The way the parameter high and low values are chosen may lead to microarchitecture configurations that are technically unrealistic or even infeasible. In other words, the various microarchitecture configurations in a Plackett and Burman experiment are corner cases in the microarchitecture design space.

The next step in the procedure is to simulate these microarchitecture configurations, collect performance numbers, and calculate the effect that each parameter has on the variation observed in the performance numbers. The latter is done by multiplying the performance number for each configuration with its value (+1 or −1) and by, subsequently, adding these products across all configurations. For example, the effect for parameter A is computed as:

$$effect_A = (1 \times 56) + (-1 \times 69) + (-1 \times 15) + \ldots + (-1 \times 67) + (1 \times 2) = -31.$$

Similarly, one can compute the effect of pairwise effects by multiplying the performance number for each configuration with the product of the parameters' values , and adding these products across all configurations. For example, the interaction effect between A and C is computed as:

$$effect_{AC} = ((1 \times 1) \times 56) + ((-1 \times 1) \times 69) + \ldots + ((1 \times 1) \times 2) = 83.$$

After having computed the effect of each parameter, the effects (including the interaction effects) can be ordered to determine their relative impact. The sign of the effect is meaningless, only the magnitude is. An effect with a higher ranking is more of a performance bottleneck than a lower ranked effect. For the example data in Table 3.4, the most significant parameter is B (with an effect of -129).

By running a Plackett and Burman experiment on a variety of benchmarks, one can compare the benchmarks against each other. In particular, the Plackett and Burman experiment yields a ranking of the most significant performance bottlenecks. Comparing these rankings across benchmarks provides a way to assess whether the benchmarks stress similar performance bottlenecks, i.e., if the top N ranked performance bottlenecks and their relative ranking is about the same for two benchmarks, one can conclude that both benchmarks exhibit similar behavior.

Yi et al. [197] compare the PCA-based and PB-based methodologies against each other. The end conclusion is that both methodologies are equally accurate in terms of how well they can identify a reduced workload. Both methods can reduce the size of the workload by a factor of 3 while incurring an error (difference in IPC for the reduced workload compared to the reference workload across a number of processor architectures) of less than 5%. In terms of computational efficiency, the PCA-based method was found to be more efficient than the PB-based approach. Collecting the PCA program characteristics was done more efficiently than running the detailed cycle-accurate simulations needed for the PB method.

3.4   LIMITATIONS AND DISCUSSION

Both the PCA-based as well as the PB-based workload design methodologies share a common pitfall, namely the reduced workload may not capture all the behaviors of the broader set of applications. In other words, there may exist important behaviors that the reduced workload does not cover. The fundamental reason is that the reduced workload is representative with respect to the reference workload from which the reduced workload is derived. The potential pitfall is that if the reference workload is non-representative with respect to the workload space, the reduced workload may not be representative for the workload space either.

Along the same line, the reduced workload is representative for the reference workload with respect to the input that was given to the workload reduction method. In particular, a PCA-based method considers a set of program characteristics to gauge behavioral similarity across benchmarks, and, as a result, evaluating a microarchitecture feature that is not captured by any of the program characteristics during workload reduction may be misleading. Similarly, the PB-based method considers a (limited) number of microarchitecture configurations during workload reduction; a completely different microarchitecture than the ones considered during workload reduction may yield potentially different performance numbers for the reduced workload than for the reference workload.

One concrete example to illustrate this pitfall is value prediction [127], which is a microarchitectural technique that predicts and speculates on the outcome of instructions. Assume that an architect wants to evaluate the performance potential of value prediction, and assume that the reduced workload was selected based on a set program characteristics (for the PCA-based method) or microarchitecture configurations (for the PB-based method) that do not capture any notion of value locality and predictability. Then the reduced workload may potentially lead to a misleading conclusion, i.e., the value predictability may be different for the reduced workload than for the reference workload, for the simple reason that the notion of value locality and predictability was not taken into account during workload reduction.

This pitfall may not be a major issue in practice though because microarchitectures typically do not change radically from one generation to the next. The transition from one generation to the next is typically smooth, with small improvements accumulating to significant performance differences over time. Nevertheless, architects should be aware of this pitfall and should, therefore, not only focus on the reduced workload during performance evaluation. It is important to evaluate the representativeness of the reduced workload with respect to the reference workload from time to time and revisit the reduced workload if needed. In spite of having to revisit the representativeness of the reduced workload, substantial simulation time savings will still be realized because (most of) the experiments are done using a limited number of benchmarks and not the entire reference workload.

CHAPTER 4

Analytical Performance Modeling

Analytical performance modeling is an important performance evaluation method that has gained increased interest over the past few years. In comparison to the prevalent approach of simulation (which we will discuss in subsequent chapters), analytical modeling may be less accurate, yet it is multiple orders of magnitude faster than simulation: a performance estimate is obtained almost instantaneously — it is a matter of computing a limited number of formulas which is done in seconds or minutes at most. Simulation, on the other hand, can easily take hours, days, or even weeks.

Because of its great speed advantage, analytical modeling enables exploring large design spaces very quickly, which makes it a useful tool in early stages of the design cycle and even allows for exploring very large design spaces that are intractable to explore through simulation. In other words, analytical modeling can be used to quickly identify a region of interest that is later explored in more detail through simulation. One example that illustrates the power of analytical modeling for exploring large design spaces is a study done by Lee and Brooks [120], which explores the potential of adaptive miroarchitectures while varying both the adaptibility of the microarchitecture and the time granularity for adaptation — this is a study that would have been infeasible through detailed cycle-accurate simulation.

In addition, analytical modeling provides more fundamental insight. Although simulation provides valuable insight as well, it requires many simulations to understand performance sensitivity to design parameters. In contrast, the sensitivity may be apparent from the formula itself in analytical modeling. As an example, Hill and Marty extend Amdahl's law towards multicore processors [82]. They augment Amdahl's law with a simple hardware cost model, and they explore the impact of symmetric (homogeneous), asymmetric (heterogeneous) and dynamic multicore processing. In spite of its simplicity, it provides fundamental insight and reveals various important consequences for the multicore era.

4.1 EMPIRICAL VERSUS MECHANISTIC MODELING

In this chapter, we classify recent work in analytical performance modeling in three major categories, empirical modeling, mechanistic modeling and hybrid empirical-mechanistic modeling. Mechanistic modeling builds a performance model based on first principles, i.e., the performance model is built in a bottom-up fashion starting from a basic understanding of the mechanics of the underlying system. Mechanistic modeling can be viewed of as 'white-box' performance modeling, and its key feature is to provide fundamental insight and, potentially, generate new knowledge. Empirical modeling, on the other hand, builds a performance model through a 'black-box' approach. Empirical modeling typically leverages statistical inference and machine learning techniques such as regression modeling or neural networks to automatically learn a performance model from training data. While inferring a performance model is easier through empirical modeling because of the complexity of the underlying system, it typically provides less insight than mechanistic modeling. Hybrid mechanistic-empirical modeling occupies the middle ground between mechanistic and empirical modeling, and it could be viewed of as 'gray-box' modeling. Hybrid mechanistic-empirical modeling starts from a generic performance formula that is derived from insights in the underlying system; however, this formula includes a number of unknown parameters. These unknown parameters are then inferred through fitting (e.g., regression), similarly to empirical modeling. The motivation for hybrid mechanistic-empirical modeling is that it provides insight (which it inherits from mechanistic modeling) while easing the construction of the performance model (which it inherits from empirical modeling).

Although we make a distinction between empirical and mechanistic modeling, there is no purely empirical or mechanistic model [145]. A mechanistic model always includes some form of empiricism, for example, in the modeling assumptions and approximations. Likewise, an empirical model always includes a mechanistic component, for example, in the list of inputs to the model — the list of model inputs is constructed based on some understanding of the underlying system. As a result, the distinction between empirical and mechanistic models is relative, and we base our classification on the predominance of the empirical versus mechanistic component in the model. We now describe the three modeling approaches in more detail.

4.2   EMPIRICAL MODELING

While empirical modeling allows users to embed domain knowledge into the model, effective models might still be constructed without such prior knowledge. This flexibility might explain the recent popularity of this modeling technique. Different research groups have proposed different approaches to empirical modeling, which we revisit here.

4.2.1   LINEAR REGRESSION

Linear regression is a widely used empirical modeling approach which relates a response variable to a number of input parameters. Joseph et al. [97] apply this technique to processor performance modeling and build linear regression models that relate micro-architectural parameters (along with some of their interactions) to overall processor performance. Joseph et al. only use linear regression to test design parameters for significance, i.e., they do not use linear regression for predictive modeling. In that sense, linear regression is similar to the PCA and Plackett-Burman approaches discussed in the previous chapter. The more advanced regression techniques, non-linear and spline-based regression, which we discuss next, have been applied successfully for predictive modeling. (And because the more advanced regression methods extend upon linear regression, we describe linear regression here.)

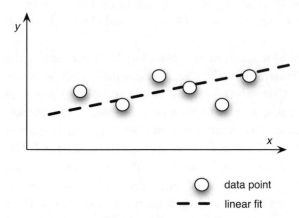

Figure 4.1: Linear regression.

The simplest form of linear regression is

$$y = \beta_0 + \sum_{i=1}^{n} \beta_i x_i + \epsilon, \tag{4.1}$$

with y the dependent variable (also called the response variable), x_i the independent variables (also called the input variables), and ϵ the error term due to lack of fit. β_0 is the intercept with the y-axis and the β_i coefficients are the regression coefficients. The β_i coefficients represent the expected change in the response variable y per unit of change in the input variable x_i; in other words, a regression coefficient represents the significance of its respective input variable. A linear regression model could potentially relate performance (response variable) to a set of microarchitecture parameters (input variables); the latter could be processor width, pipeline depth, cache size, cache latency, etc. In other words, linear regression tries to find the best possible linear fit for a number of data points, as illustrated in Figure 4.1.

This simple linear regression model assumes that the input variables are independent of each other, i.e., the effect of variable x_i on the response y does not depend on the value of x_j, $j \neq i$. In many cases, this is not an accurate assumption, especially in computer architecture. For example, the effect on performance of making the processor pipeline deeper depends on the configuration of the memory hierarchy. A more aggressive memory hierarchy reduces cache miss rates, which reduces average memory access times and increases pipelining advantages. Therefore, it is possible to consider interaction terms in the regression model:

$$y = \beta_0 + \sum_{i=1}^{n} \beta_i x_i + \sum_{i=1}^{n} \sum_{j=i+1}^{n} \beta_{i,j} x_i x_j + \epsilon. \tag{4.2}$$

This particular regression model only includes, so called, two-factor interactions, i.e., pairwise interactions between two input variables only; however, this can be trivially extended towards higher order interactions.

The goal for applying regression modeling to performance modeling is to understand the effect of the important microarchitecture parameters and their interactions on overall processor performance. Joseph et al. [97] present such an approach and select a number of microarchitecture parameters such as pipeline depth, processor width, reorder buffer size, cache sizes and latencies, etc., along with a selected number of interactions. They then run a number of simulations while varying the microarchitecture parameters and fit the simulation results to the regression model. The method of least squares is commonly used to find the best fitting model that minimizes the sum of squared deviations between the predicted response variable (through the model) and observed response variable (through simulation). The fitting is done such that the error term is as small as possible. The end result is an estimate for each of the regression coefficients. The magnitude and sign of the regression coefficients represent the relative importance and impact of the respective microarchitecture parameters on overall performance.

There are a number of issues one has to deal with when building a regression model. For one, the architect needs to select the set of microarchitecture input parameters, which has an impact on both accuracy and the number of simulations needed to build the model. Insignificant parameters only increase model building time without contributing to accuracy. On the other hand, crucial parameters that are not included in the model will likely lead to an inaccurate model. Second, the value ranges that need to be set for each variable during model construction depends on the purpose of the experiment. Typically, for design space exploration purposes, it is advised to take values that are slightly outside the expected parameter range — this is to cover the design space well [196].

Table 4.1 shows the most significant variables and interactions obtained in one of the experiments done by Joseph et al. They consider six microarchitecture parameters and their interactions as the input variables, and they consider IPC as their response performance metric. Some parameters and interactions are clearly more significant than others, i.e., their respective regression coefficients have a large magnitude: pipeline depth and reorder buffer size as well as its interaction are significant, much more significant than L2 cache size. As illustrated in this case study, the regression coefficients can be both positive and negative, which complicates gaining insight. In particular, Table 4.1 suggests that IPC decreases with increasing reorder buffer size because the regression coefficient is negative. This obviously makes no sense. The negative regression coefficient is compensated for by the positive interaction terms between reorder buffer size and pipeline depth, and reorder buffer size and issue buffer size. In other words, increasing the reorder buffer size will increase the interaction terms more than the individual variable so that IPC would indeed increase with reorder buffer size.

Table 4.1: Example illustrating the output of a linear regression experiment done by Joseph et al. [97].

Intercept	1.230
Pipeline depth	−0.566
Reorder buffer size	−0.480
Pipeline depth × reorder buffer size	0.378
Issue queue size	−0.347
Reorder buffer size × issue queue size	0.289
Pipeline depth × issue queue size	0.274
Pipeline depth × reorder buffer size × issue queue size	−0.219

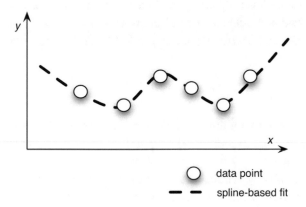

data point

spline-based fit

Figure 4.2: Spline-based regression.

4.2.2 NON-LINEAR AND SPLINE-BASED REGRESSION

Basic linear regression as described in the previous section assumes that the response variable behaves linearly with its input variables. This assumption is often too restrictive. There exist several techniques for capturing non-linearity.

The most simple approach is to transform the input variables or response variable or both. Typical transformations are square root, logarithmic, power, etc. The idea is that such transformations make the response more linear and thus easier to fit. For example, instruction throughput (IPC) is known to relate to reorder buffer size following an approximate square root relation [138; 166], so it makes sense to take the square root of the reorder buffer variable in an IPC model in order to have a better fit. The limitation is that the transformation is applied to an input variable's entire range, and thus a good fit in one region may unduly affect the fit in another region.

Lee and Brooks [119] advocate spline-based regression modeling in order to capture non-linearity. A spline function is a piecewise polynomial used in curve fitting. A spline function is partitioned in a number of intervals with different continuous polynomials. The endpoints for the polynomials are called knots. Denoting the knots' x-values as x_i and their y-values as y_i, the spline is then defined as

$$S(x) = \begin{cases} S_0(x) & x \in [x_0, x_1] \\ S_1(x) & x \in [x_1, x_2] \\ \dots & \\ S_n(x) & x \in [x_{n-1}, x_n], \end{cases} \tag{4.3}$$

with each $S_i(x)$ a polynomial. Higher-order polynomials typically lead to better fits. Lee and Brooks use cubic splines which have the nice property that the resulting curve is smooth because the first and second derivatives of the function are forced to agree at the knots. Restricted cubic splines constrain the function to be linear in the tails; see Figure 4.2 for an example restricted cubic spline. Lee and Brooks successfully leverage spline-based regression modeling to build multiprocessor performance models [122], characterize the roughness of the architecture design space [123], and explore the huge design space of adaptive processors [120].

4.2.3 NEURAL NETWORKS

Artificial neural networks are an alternative approach to building empirical models. Neural networks are machine learning models that automatically learn to predict (a) target(s) from a set of inputs. The target could be performance and/or power or any other metric of interest, and the inputs could be microarchitecture parameters. Neural networks could be viewed of as a generalized non-linear regression model. Several groups have explored the idea of using neural networks to build performance models, see for example Ipek et al. [89], Dubach et al. [41] and Joseph et al. [98]. Lee et al. [121] compare spline-based regression modeling against artificial neural networks and conclude that both approaches are equally accurate; regression modeling provides better statistical understanding while neural networks offer greater automation.

Figure 4.3(a) shows the basic organization of a fully connected feed-forward neural network. The network consists of one input layer and one output layer, and one or more hidden layers. The input layer collects the inputs to the model, and the output layer provides the model's predictions. Data flows from the inputs to the outputs. Each node is connected to all nodes from the previous layer. Each edge has a weight and each node in the hidden and output layers computes the weighted sum of its inputs. The nodes in the hidden layer apply the weighted sum of its inputs to a so called activation function, see also Figure 4.3(b). A commonly used activation function is the sigmoid function, which is a mathematical function having an 'S' shape with two horizontal asymptotes.

Training an artificial neural network basically boils down to a search problem that aims at finding the weights, such that the error between the network's predictions and the corresponding measurements is minimized. Training the neural network is fairly similar to inferring a regression model: the network's edge weights are adjusted to minimize the squared error between the simulation results and the model predictions. During training, examples are repeatedly presented at the inputs, differences between network outputs and target values are calculated, and weights are updated by

(a) organization of a fully connected feed-forward neural network

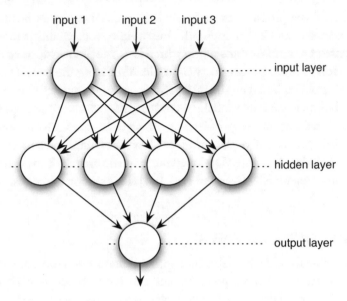

(b) an individual node

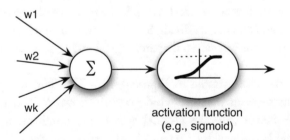

Figure 4.3: Neural networks: (a) architecture of a fully connected feed-forward network, and (b) architecture for an individual node.

taking a small step in the direction of steepest decrease in error. This is typically done through a well-known procedure called backpropagation.

A limitation for empirical modeling, both neural networks and regression modeling, is that it requires a number of simulations to infer the model. This number of simulations typically varies between a couple hundreds to a few thousands of simulations. Although this is time consuming to do, it is a one-time cost. Once the simulations are run and once the model is built, making performance predictions is done instantaneously.

4.3   MECHANISTIC MODELING: INTERVAL MODELING

Mechanistic modeling takes a different approach: it starts from a basic understanding of the underlying system from which a performance model is then inferred. One could view mechanistic modeling as a bottom-up approach, in contrast to empirical modeling which is as a top-down approach.

Building mechanistic models for early processors was simple. Measuring the instruction mix and adding a constant cost per instruction based on the instruction's execution latency and memory access latency was sufficient to build an accurate model. In contrast, contemporary processors are much more complicated and implement various ways of latency hiding techniques (instruction-level parallelism, memory-level parallelism, speculative execution, etc.), which complicates mechanistic performance modeling. Interval analysis is a recently developed mechanistic model for contemporary superscalar out-of-order processors presented by Eyerman et al. [64]. This section describes interval modeling in more detail.

4.3.1   INTERVAL MODEL FUNDAMENTALS

Figure 4.4(a) illustrates the fundamentals of the model: under optimal conditions, i.e., in the absence of miss events, a balanced superscalar out-of-order processor sustains instructions per cycle (IPC) performance roughly equal to its dispatch width D — dispatch refers to the movement of instructions from the front-end pipeline into the reorder and issue buffers of a superscalar out-of-order processor. However, when a miss event occurs, the dispatching of useful instructions eventually stops. There is then a period when no useful instructions are dispatched, lasting until the miss event is resolved, and then instructions once again begin flowing. Miss events divide execution time into intervals, which begin and end at the points where instructions just begin dispatching following recovery from the preceding miss event.

Each interval consists of two parts, as illustrated in Figure 4.4(b). The first part performs useful work in terms of dispatching instructions into the window: if there are N instructions in a given interval (interval length of N) then it will take $\lceil N/D \rceil$ cycles to dispatch them into the window. The second part of the interval is the penalty part and is dependent on the type of miss event. The exact mechanisms which cause the processor to stop and re-start dispatching instructions into the window, and the timing with respect to the occurrence of the miss event are dependent on the type of miss event, so each type of miss event must be analyzed separately.

4.3.2   MODELING I-CACHE AND I-TLB MISSES

L1 I-cache misses, L2 instruction cache misses and I-TLB misses are the easiest miss events to handle, see also Figure 4.5. At the beginning of the interval, instructions begin to fill the window at a rate equal to the maximum dispatch width. Then, at some point, an instruction cache miss occurs. Fetching stops while the cache miss is resolved, but the front-end pipeline is drained, i.e., the instructions already in the front-end pipeline are dispatched into the reorder and issue buffers, and then dispatch stops. After a delay for handling the I-cache miss, the pipeline begins to re-fill

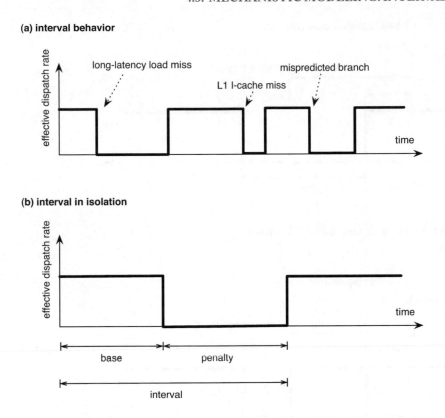

Figure 4.4: Interval behavior: (a) overall execution can be split up in intervals; (b) an interval consists of a base part where useful work gets done and a penalty part.

and dispatch is resumed. The front-end pipeline re-fill time is the same as the drain time — they offset each other. Hence, the penalty for an I-cache (and I-TLB) miss is its miss delay.

4.3.3   MODELING BRANCH MISPREDICTIONS

Figure 4.6 shows the timing for a branch misprediction interval. At the beginning of the interval, instructions are dispatched, until, at some point, the mispredicted branch is dispatched. Although wrong-path instructions continue to be dispatched (as displayed with the dashed line in Figure 4.6), dispatch of useful instructions stops at that point. Then, useful dispatch does not resume until the mispredicted branch is resolved, the pipeline is flushed, and the instruction front-end pipeline is re-filled with correct-path instructions.

The overall performance penalty due to a branch misprediction thus equals the difference between the time the mispredicted branch enters the window and the time the first correct-path instruction enters the window following discovery of the misprediction. In other words, the overall

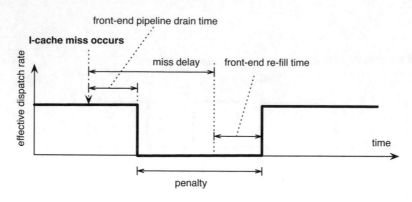

Figure 4.5: Interval behavior of an I-cache/TLB miss.

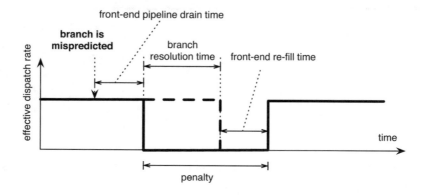

Figure 4.6: Interval behavior of a mispredicted branch.

performance penalty equals the branch resolution time, i.e., the time between the mispredicted branch entering the window and the branch being resolved, plus the front-end pipeline depth. Eyerman et al. [65] found that the mispredicted branch often is the last instruction to be executed; and hence, the branch resolution time can be approximated by the 'window drain time', or the number of cycles needed to empty a reorder buffer with a given number of instructions. For many programs, the branch resolution time is the main contributor to the overall branch misprediction penalty (not the pipeline re-fill time). And this branch resolution time is a function of the dependence structure of the instructions in the window, i.e., the longer the dependence chain and the execution latency of the instructions leading to the mispredicted branch, the longer the branch resolution time [65].

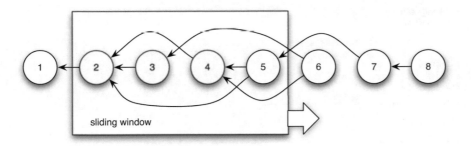

Figure 4.7: ILP model: window cannot slide any faster than determined by the critical path.

4.3.4   MODELING SHORT BACK-END MISS EVENTS USING LITTLE'S LAW

An L1 D-cache miss is considered to be a 'short' back-end miss event and is modeled as if it is an instruction that is serviced by a long-latency functional unit, similar to a multiply or a divide. In other words, it is assumed that the miss latency can be hidden by out-of-order execution, and this assumption is incorporated into the model's definition of a balanced processor design. In particular, the ILP model for balanced processor designs includes L1 D-cache miss latencies as part of the average instruction latency when balancing reorder buffer size and issue width in the absence of miss events. The ILP model is based on the notion of a window (of a size equal to the reorder buffer size) that slides across the dynamic instruction stream, see Figure 4.7. This sliding window computes the critical path length or the longest data dependence chain of instructions (including their execution latencies) in the window. Intuitively, the window cannot slide any faster than the processor can issue the instructions belonging to the critical path.

The ILP model is based on Little's law, which states that the throughput through a system equals the number of elements in the system divided by the average time spent for each element in the system. When applied to the current context, Little's law states that the IPC that can be achieved over a window of instructions equals the number of instructions in the reorder buffer (window) W divided by the average number of cycles ℓ an instruction spends in the reorder buffer (between dispatch and retirement):

$$IPC = \frac{W}{\ell}. \tag{4.4}$$

The total time an instruction spends in the reorder buffer depends on the instruction's execution latency and the dependency chain leading to the instruction, i.e., the critical path determines the achievable instruction-level parallelism (ILP). This ILP model has important implications. Knowing the critical path length as a function of window size, enables computing the achievable steady-state IPC for each window size. Or, in other words, the ILP model states which reorder buffer size is needed in order to have a balanced design and achieve an IPC close to the designed processor width in the absence of miss events.

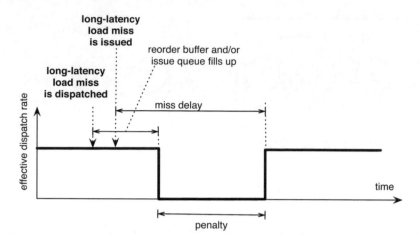

Figure 4.8: Interval behavior of an isolated long-latency load.

The ILP model is only one example illustrating the utility of Little's law — Little's law is widely applicable in systems research and computer architecture. It can be applied as long as the three parameters (throughput, number of elements in the system, latency of each element) are long-term (steady-state) averages of a stable system. There are multiple examples of how one could use Little's law in computer architecture. One such example relates to computing the number of physical registers needed in an out-of-order processor. Knowing the target IPC and the average time between acquiring and releasing a physical register, one can compute the required number of physical registers. Another example relates to computing the average latency of a packet in a network. Tracking the latency for each packet may be complex to implement in an FPGA-based simulator in an efficient way. However, Little's law offers an easy solution: it suffices to count the number of packets in the network and the injection rate during steady-state, and compute the average packet latency using Little's law.

4.3.5   MODELING LONG BACK-END MISS EVENTS

When a long data cache miss occurs, i.e., from the L2 cache to main memory, the memory delay is typically quite large — on the order of hundreds of cycles. Similar behavior is observed for D-TLB misses.

On an isolated long data cache miss due to a load, the reorder buffer fills because the load blocks the reorder buffer head [102], and then dispatch stalls, see Figure 4.8. After the miss data returns from memory, the load can be executed and committed, which unblocks the reorder buffer, and as a result, instruction dispatch resumes. The total long data cache miss penalty equals the time between the ROB fill, and the time data returns from memory. The penalty for an isolated long back-end miss thus equals the main memory access latency minus the number of cycles where useful

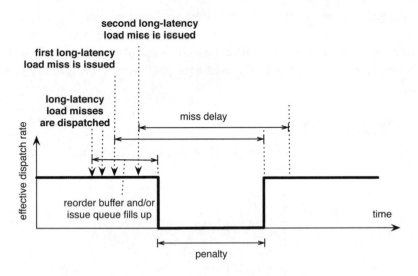

Figure 4.9: Interval behavior of two independent overlapping long-latency loads.

instructions are dispatched under the long-latency miss. These useful instructions are dispatched between the time the long-latency load dispatches and the time the ROB blocks after the long-latency load reaches its head; this is the time it takes to fill the entire ROB minus the time it takes for the load to issue after it has been dispatched — this is the amount of useful work done underneath the memory access. Given that this is typically much smaller than the memory access latency, the penalty for an isolated miss is assumed to equal the memory access latency.

For multiple long back-end misses that are independent of each other and that make it in the reorder buffer at the same time, the penalties overlap [31; 76; 102; 103] — this is referred to as memory-level parallelism (MLP). This is illustrated in Figure 4.9. After the first load receives its data and unblocks the ROB, S more instructions dispatch before the ROB blocks for the second load, and the time to do so, S/D, offsets an equal amount of the second load's miss penalty. This generalizes to any number of overlapping misses, so the penalty for a burst of independent long-latency back-end misses equals the penalty for an isolated long-latency load.

4.3.6 MISS EVENT OVERLAPS

So far, we considered the various miss event types in isolation. However, the miss events may interact with each other. The interaction between front-end miss events (branch mispredictions and I-cache/TLB misses) is limited because they serially disrupt the flow of instructions and thus their penalties serialize. Long-latency back-end miss events interact frequently and have a large impact on overall performance as discussed in the previous section, but these can be modeled fairly easily by counting the number of independent long-latency back-end misses that occur within an instruction sequence less than or equal to the reorder buffer size. The interactions between front-end miss events and long-latency back-end miss events are more complex because front-end miss events can overlap back-end misses; however, these second-order effects do not occur often (at most 5% of the total run time according to the experiments done by Eyerman et al. [64]), which is why interval modeling simply ignores them.

4.3.7 THE OVERALL MODEL

When put together, the model estimates the total execution time in the number of cycles C on a balanced processor as:

$$
\begin{aligned}
C \;=\; & \sum \lceil \frac{N_k}{D} \rceil \\
& + m_{iL1} \cdot c_{iL1} + m_{iL2} \cdot c_{L2} \\
& + m_{br} \cdot (c_{dr} + c_{fe}) \\
& + m_{dL2}^{*}(W) \cdot c_{L2}.
\end{aligned}
$$

The various parameters in the model are summarized in Table 4.2. The first line of the model computes the total number of cycles needed to dispatch all the intervals. Note there is an inherent dispatch inefficiency because the interval length N_k is not always an integer multiple of the processor dispatch width D, i.e., fewer instructions may be dispatched at the trailer of an interval than the designed dispatch width, simply because there are too few instructions to the next interval to fill the entire width of the processor's front-end pipeline. The subsequent lines in Equation 4.5 represent I-cache misses, branch mispredictions and long back-end misses, respectively. (The TLB misses are not shown here to increase the formula's readability.) The I-cache miss cycle component is the number of I-cache misses times their penalty. The branch misprediction cycle component equals the number of mispredicted branches times their penalty, the window drain time plus the front-end pipeline depth. Finally, the long back-end miss cycle component is computed as the number of non-overlapping misses times the memory access latency.

4.3.8 INPUT PARAMETERS TO THE MODEL

The model has two sets of program characteristics. The first set of program characteristics are related to a program's locality behavior and include the miss rates and interval lengths for the various miss events. The second program characteristic relates to the branch resolution time which is approximated by the window drain time c_{dr}. Note that these two sets of program characteristics are the only program characteristics needed by the model; all the other parameters are microarchitecture-related. The locality metrics can be obtained through modeling or through specialized trace-driven simulation (i.e., cache, TLB and branch predictor simulation). The window drain time is estimated through the ILP model.

Table 4.2: Parameters to the interval model.	
C	Total number of cycles
N_k	Interval length for interval k
D	Designed processor dispatch width
m_{iL1} and m_{iL2}	Number of L1 and L2 I-cache misses
c_{iL1} and c_{iL2}	L1 and L2 I-cache miss delay
m_{br}	Number of mispredicted branches
c_{dr}	Window drain time
c_{fe}	Front-end pipeline depth
$m^*_{dL2}(W)$	Number of non-overlapping L2 cache load misses for a given reorder buffer size W
c_{L2}	L2 D-cache miss delay

4.3.9   PREDECESSORS TO INTERVAL MODELING

A number of primary efforts led to the development of the interval model. Michaud et al. [138] build a mechanistic model of the instruction window and issue mechanism in out-of-order processors for gaining insight in the impact of instruction fetch bandwidth on overall performance. Karkhanis and Smith [103; 104] extend this simple mechanistic model to build a complete performance model that assumes sustained steady-state performance punctuated by miss events. Taha and Wills [180] propose a mechanistic model that breaks up the execution into so called macro blocks, separated by miss events. These earlier models focus on the issue stage of a superscalar out-of-order processor — issue refers to selecting instructions for execution on the functional units. The interval model as described here (see also [64] for more details) focuses on dispatch rather than issue, which makes the model more elegant. Also, the ILP model in [64] eliminates the need for extensive micro-architecture simulations during model construction, which the prior works needed to determine 'steady-state' performance in the absence of miss events.

4.3.10   FOLLOW-ON WORK

Eyerman et al. [63] use the interval model to provide the necessary insights to develop a hardware performance counter architecture that can compute accurate CPI components and CPI stacks in superscalar out-of-order architectures. CPI stacks are stacked bars with the base cycle component typically shown at the bottom and the other CPI components stacked on top of it. CPI stacks are useful for guiding software and system optimization because it visualizes where the cycles have gone. Genbrugge et al. [73] replace the core-level cycle-accurate simulation models in a multicore simulator by the interval model; the interval model then estimates the progress for each instruction in the core's pipeline. The key benefits are that the interval model is easier to implement than a cycle-accurate core simulator, and in addition, it runs substantially faster. Karkhanis and Smith [104] use the interval model to explore the processor design space automatically and identify processor configurations that represent Pareto-optimal design points with respect to performance, energy and chip area for a particular application or set of applications. Chen and Aamodt [27] extend the interval model by proposing ways to include hardware prefetching and account for a limited number of miss status handling registers (MSHRs). Hong and Kim [84] present a first-order model for GPUs which shares some commonalities with the interval model described here.

4.3.11 MULTIPROCESSOR MODELING

The interval model discussed so far focuses on modeling individual cores, and it does not target shared memory multiprocessor nor chip-multiprocessors. Much older work by Sorin et al. [175] proposed an analytical model for shared memory multiprocessors. This model assumes a black-box model for the individual processors — a processor is considered to generate requests to the memory system and intermittently block after a (dynamically changing) number of requests — and models the memory system through mean value analysis, which is a white-box model. (This model illustrates what we stated earlier, namely, a model is not purely mechanistic or empirical, and a mechanistic model may involve some form of empiricism.) The models views the memory system as a system consisting of queues (e.g., memory bus, DRAM modules, directories, network interfaces) and delay centers (e.g., switches in the interconnection network). Mean value analysis is a technique within queueing theory that estimates the expected queue lengths in a closed system of queues. The model then basically estimates the total time for each request to the memory system by adding the request's mean residence time in each of the resources that it visits (e.g., processor, network interface at the sender, network, network interface at the receiver, bus at the receiver, memory and directory at the receiver side).

4.4   HYBRID MECHANISTIC-EMPIRICAL MODELING

Hybrid mechanistic-empirical modeling combines the best of both worlds: it combines the insight from mechanistic modeling with the ease of model development from empirical modeling. Hybrid mechanistic-empirical modeling starts from a performance model that is based on some insight from the underlying system; however, there are a number of unknowns. These unknowns are parameters that are subsequently fit using training data, similar to what is done in empirical modeling.

Hartstein and Puzak [78] propose a hybrid mechanistic-empirical model for studying optimum pipeline depth. The model is parameterized with a number of parameters that are fit through detailed, cycle-accurate micro-architecture simulations. Hartstein and Puzak divide the total execution time in busy time T_{BZ} and non-busy time T_{NBZ}. The busy time refers to the time that the processor is doing useful work, i.e., instructions are issued; the non-busy time refers to the time that execution is stalled due to miss events. Hartstein and Puzak derive that the total execution time equals busy time plus non-busy time:

$$T = N_{total}/\alpha \cdot (t_o + t_p/p) + N_H \cdot \gamma \cdot (t_o \cdot p + t_p) \qquad (4.5)$$

with N_{total} the total number of dynamically executed instructions and N_H the number of hazards or miss events; t_o the latch overhead for a given technology, t_p the total logic (and wire) delay of a processor pipeline, and p the number of pipeline stages. The α and γ parameters are empirically derived by fitting the model to data generated with detailed simulation.

CHAPTER 5

Simulation

Simulation is the prevalent and de facto performance evaluation method in computer architecture. There are several reasons for its widespread use. Analytical models, in spite of the fact that they are extremely fast to evaluate and in spite of the deep insight that they provide, incur too much inaccuracy for many of the design decisions that an architect needs to make. One could argue that analytical modeling is valuable for making high-level design decisions and identifying regions of interest in the huge design space. However, small performance variations across design alternatives are harder to evaluate using analytical models. At the other end of the spectrum, hardware prototypes, although they are extremely accurate, are too time-consuming and costly to develop.

A simulator is a software performance model of a processor architecture. The processor architecture that is modeled in the simulator is called the target architecture; running the simulator on a host architecture, i.e., a physical machine, then yields performance results. Simulation has the important advantage that development is relatively cheap compared to building hardware prototypes, and it is typically much more accurate than analytical models. Moreover, the simulator is flexible and easily parameterizable which allows for exploring the architecture design space — a property of primary importance to computer architects designing a microprocessor and researchers evaluating a novel idea. For example, evaluating the impact of cache size, latency, processor width, branch predictor configuration is easily done through parameterization, i.e., by changing some of the simulator's parameters and running a simulation with a variety of benchmarks, one can evaluate what the impact is of an architecture feature. Simulation even enables evaluating (very) different architectures than the ones in use today.

5.1 THE COMPUTER ARCHITECT'S TOOLBOX

There exist many flavors of simulation, each representing a different trade-off in accuracy, evaluation time, development time and coverage. Accuracy refers to the fidelity of the simulation model with respect to real hardware, i.e., how accurate is the simulation model compared to the real hardware that it models. Evaluation time refers to how long it takes to run a simulation. Development time refers to how long it takes to develop the simulator. Finally, coverage relates to what fraction of the design space a simulator can explore, e.g., a cache simulator can only be used to evaluate cache performance and not overall processor performance.

This simulation trade-off can be represented as a diamond, see Figure 5.1. (The trade-off is often represented as a triangle displaying accuracy, evaluation time and development time; a diamond illustrates a fourth crucial dimension.) Each simulation approach (or a modeling effort in

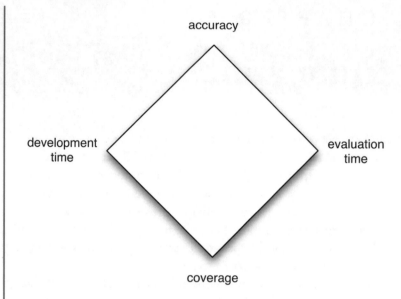

Figure 5.1: Simulation diamond illustrates the trade-offs in simulator accuracy, coverage, development time and evaluation time.

general) can be characterized along these four dimensions. These dimensions are not independent of each other, and, in fact, are contradictory. For example, more faithful modeling with respect to real hardware by modeling additional features, i.e., increasing the simulator's coverage, is going to increase accuracy, but it is also likely to increase the simulator's development and evaluation time — the simulator will be more complex to build, and because of its increased complexity, it will also run slower, and thus simulation will take longer. In contrast, a simulator that only models a component of the entire system, e.g., a branch predictor or cache, has limited coverage with respect to the entire system; nevertheless, it is extremely valuable because its accuracy is good for the component under study while being relatively simple (limited development time) and fast (short evaluation time).

The following sections describe several commonly used simulation techniques in the computer architect's toolbox, each representing a different trade-off in accuracy, coverage, development time and evaluation time. We will refer to Table 5.1 throughout the remainder of chapter; it summarizes the different simulation techniques along the four dimensions.

5.2   FUNCTIONAL SIMULATION

Functional simulation models only the functional characteristics of an instruction set architecture (ISA) — it does not provide any timing estimates. That is, instructions are simulated one at a time, by taking input values and computing output values. Therefore, functional simulators are also called instruction-set emulators. These tools are typically most useful for validating the correctness of a de-

Table 5.1: Comparing functional simulation, instrumentation, specialized cache and predictor simulation, full trace-driven simulation and full execution-driven simulation in terms of model development time, evaluation time, accuracy in predicting overall performance, and level of detail or coverage.

	Development time	Evaluation time	Accuracy	Coverage
functional simulation	excellent	good	poor	poor
instrumentation	excellent	very good	poor	poor
specialized cache and predictor simulation	good	good	good	limited
full trace-driven simulation	poor	poor	very good	excellent
full execution-driven simulation	very poor	very poor	excellent	excellent

sign rather than evaluating its performance characteristics. Consequently, the accuracy and coverage with respect to performance and implementation detail are not applicable. However, development time is rated as excellent because a functional simulator is usually already present at the time a hardware development project is undertaken (unless the processor implements a brand new instruction set). Functional simulators have a very long lifetime that can span many development projects. Evaluation time is good because no microarchitecture features need to be modeled. Example functional simulator are SimpleScalar's sim-safe and sim-fast [7].

From a computer architect's perspective, functional simulation is most useful because it can generate instruction and address traces. A trace is the functionally correct sequence of instructions and/or addresses that a benchmark program produces. These traces can be used as inputs to other simulation tools — so called (specialized) trace-driven simulators.

5.2.1 ALTERNATIVES

An alternative to functional simulation is instrumentation, also called direct execution. Instrumentation takes a binary and adds code to it so that when running the instrumented binary on real hardware the property of interest is collected. For example, if the goal is to generate a trace of memory addresses, it suffices to instrument (i.e., add code to) each instruction referencing memory in the binary to compute and print the memory address; running the instrumented binary on native hardware then provides a trace of memory addresses. The key advantage of instrumentation compared to functional simulation is that it incurs less overhead. Instrumentation executes all the instructions natively on real hardware; in contrast, functional simulation emulates all the instructions and hence executes more host instructions per target instruction. There exist two flavors of instrumentation, static instrumentation, which instruments the binary statically, and dynamic instrumentation, which instruments the binary at run time. An example tool for static instrumentation is Atom [176], or EEL [114] (used in the Wisconsin Wind Tunnel II simulator [142]); Embra [191], Shade [34] and Pin [128] support dynamic instrumentation. A limitation of instrumentation compared to functional simulation is that the target ISA is typically the same as the host ISA, and thus an instrumentation framework is not easily portable. A dynamic binary translator that translates host ISA instructions to target ISA instructions can address this concern (as is done in the Shade framework); however, the

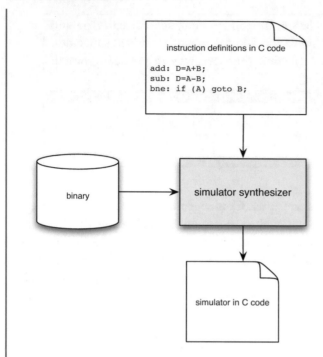

Figure 5.2: Functional simulator synthesizer proposed by Burtscher and Ganusov [21].

simulator can only run on a machine that implements the target ISA. Zippy, a static instrumentation system at Digital in the late 1980s, reads in an Alpha binary, and adds ISA emulation and modeling code in MIPS code.

An approach that combines the speed of instrumentation with the portability of functional simulation was proposed by Burtscher and Ganusov [21], see also Figure 5.2. They propose a functional simulator synthesizer which takes as input a binary executable as well as a file containing C definitions (code snippets) of all the supported instructions. The synthesizer then translates instructions in the binary to C statements. If desirable, the user can add simulation code to collect for example a trace of instructions or addresses. Compiling the synthesized C code generates the customized functional simulator.

5.2.2   OPERATING SYSTEM EFFECTS

Functional simulation is often limited to user-level code only, i.e., application and system library code, however, it does not simulate what happens upon an operating system call or interrupt. Nevertheless, in order to have a functionally correct execution, one needs to correctly emulate the system effects that affect the application code. A common approach is to ignore interrupts and emulate the effects of system calls [20]. Emulating system calls is typically done by manually identifying the input and output register and memory state to a system call and invoking the system call natively. This needs to be done for every system call, which is tedious and labor-intensive, especially if one wants to port a simulator to a new version of the operating system or a very different operating system.

Narayanasamy et al. [144] present a technique that automatically captures the side effects of operating system interactions. An instrumented binary collects for each system call executed, interrupt and DMA transfer, how it changes register state and memory state. The memory state change is only recorded if the memory location is later read by a load operation. This is done as follows. The instrumented binary maintains a user-level copy of the application's address space. A system effect (e.g., system call, interrupt, DMA transfer) will only affect the application's address space, not the user-level copy. A write by the application updates both the application's address space and the user-level copy. A read by the application verifies whether the data read in the application's address space matches the data in the user-level copy; upon a mismatch, the system knows that the application's state was changed by a system effect, and thus it knows that the load value in the application's address space needs to be logged. These state changes are stored in a so called system effect log. During functional simulation, the system effect log is read when reaching a system call, and the state change which is stored in the log is replayed, i.e., the simulated register and memory state is modified to emulate the effect of the system call. Because this process does not depend on the semantics of system calls, it is completely automatic, which eases developing and porting user-level simulators. A side-effect of this technique is that it enables deterministic simulation, i.e., the system effects are the same across runs. While this facilitates comparing design alternatives, it also comes with its pitfall, as we will discuss in Section 5.6.2.

5.3   FULL-SYSTEM SIMULATION

User-level simulation is sufficiently accurate for some workloads, e.g., SPEC CPU benchmarks spend little time executing system-level code, hence limiting the simulation to user-level code is sufficient. However, for other workloads, e.g., commercial workloads such as database servers, web servers, email servers, etc., simulating only user-level code is clearly insufficient because these workloads spend a considerable amount of time executing system-level code, and hence these workloads require full-system simulation. Also, the proliferation of multicore hardware has increased the importance of full-system simulation because multi-threaded workload performance is affected by OS scheduling decisions; not simulating the OS may lead to inaccurate performance numbers because it does not account for OS effects.

Full-system simulation refers to simulating an entire computer system such that complete software stacks can run on the simulator. The software stack includes application software as well as unmodified, commercial operating systems, so that the simulation includes I/O and OS activity next to processor and memory activity. In other words, a full-system simulation could be viewed of as a system emulator or a system virtual machine which appears to its user as virtual hardware, i.e., the user of the full-system simulator is given the illusion to run on real hardware. Well-known examples of full-system simulators are SimOS [167], Virtutech's SimICs [132], AMD's SimNow (x86 and x86-64) [12], M5 [16], Bochs (x86 and x86-64) [139], QEMU, Embra [191], and IBM's Mambo (PowerPC) [19].

The functionality provided by a full-system simulator is basically the same as for a user-level functional simulator — both provide a trace of dynamically executed instructions — the only difference being that functional simulation simulates user-level code instructions only, whereas full-system simulation simulates both user-level and system-level code. Full-system simulation thus achieves greater coverage compared to user-level simulation; however, developing a full-system simulator is far from trivial.

5.4 SPECIALIZED TRACE-DRIVEN SIMULATION

Specialized trace-driven simulation takes instruction and address traces — these traces may include user-level instructions only or may contain both user-level and system-level instructions — and simulates specific components, e.g., cache or branch predictor, of a target architecture in isolation. Performance is usually evaluated as a 'miss rate'. A number of these tools are widely available, especially for cache simulation, see for example Dinero IV [44] from the University of Wisconsin–Madison. In addition, several proposals have been made for simulating multiple cache configurations in a single simulation run [35; 83; 135; 177]. While development time and evaluation time are both good, coverage is limited because only certain components of a processor are modeled. And, while the accuracy in terms of miss rate is quite good, overall processor performance accuracy is only roughly correlated with these miss rates because many other factors come into play. Nevertheless, specialized trace-driven simulation has its place in the toolbox because it provides a way to easily evaluate specific aspects of a processor.

5.5 TRACE-DRIVEN SIMULATION

Full trace-driven simulation, or trace-driven simulation for short, takes program instruction and address traces, and feeds the full benchmark trace into a detailed microarchitecture timing simulator. A trace-driven simulator separates the functional simulation from the timing simulation. This is often useful because the functional simulation needs to be performed only once, while the detailed timing simulation is performed many times when evaluating different microarchitectures. This separation reduces evaluation time somewhat. Overall, full trace-driven simulation requires a long development time and requires long simulation run times, but both accuracy and coverage are very good.

One obvious disadvantage of this approach is the need to store the trace files, which may be huge for contemporary benchmarks and computer programs with very long run times. Although disk space is cheap these days, trace compression can be used to address this concern; several approaches have been made to computer trace compression [22; 95].

Another disadvantage for modern superscalar processors is that they predict branches and execute many instructions speculatively — speculatively executed instructions along mispredicted paths are later nullified. These nullified instructions do not show up in a trace file generated via functional simulation, although they may affect cache and/or predictor contents [11; 143]. Hence, trace-driven simulation will not accurately model the effects along mispredicted paths.

An additional limitation when simulating multi-threaded workloads is that trace-driven simulation cannot model the interaction between inter-thread ordering and the target microarchitecture. The reason is that the trace is fixed and imposes a particular ordering. However, the ordering and inter-thread dependences may be different across microarchitectures. For some studies, this effect may be limited; however, for other studies, it may be significant. All depends on the type of optimization and the workloads being evaluated. The key problem is that changes in some microarchitecture structure (e.g., branch predictor, cache, prefetcher, etc.) — this could even be small changes — may cause threads to acquire locks in a different order. This may lead to different conflict and contention behavior in shared resources (e.g., caches, memory, interconnection network, etc.), which, in its turn, may affect the inter-thread interleaving. Hence, even small changes in the microarchitecture can lead to (very) different performance numbers, and these changes may lead to big differences for particular benchmarks only; hence, a big change for one particular benchmark may not be representative of other workloads. Because trace-driven simulation simulates a single ordering, it cannot capture these effects. Moreover, a trace may reflect a particular ordering that may not even occur on the target microarchitecture.

5.6 EXECUTION-DRIVEN SIMULATION

In contrast to trace-driven simulation, execution-driven simulation combines functional with timing simulation. By doing so, it eliminates the disadvantages of trace-driven simulation: trace files do not need to be stored, speculatively executed instructions get simulated accurately, and the inter-thread ordering in multi-threaded workloads is modeled accurately. For these reasons, execution-driven simulation has become the de-facto simulation approach. Example execution-driven simulators are SimpleScalar [7], RSIM [88], Asim [57], M5 [16], GEMS [133], Flexus [190], and PTLSim [199]. Although execution-driven simulation achieves higher accuracy than trace-driven simulation, it comes at the cost of increased development time and evaluation time.

5.6.1 TAXONOMY

Mauer et al. [136] present a useful taxonomy of execution-driven simulators, see also Figure 5.3. The taxonomy reflects four different ways of how to couple the functional and timing components in order to manage simulator complexity and development time. An execution-driven simulator that tightly integrates the functional and timing components, hence called integrated execution-driven simulator (see Figure 5.3(a)), is obviously harder to develop and maintain. An integrated simulator is not flexible, is harder to extend (e.g., when evaluating a new architectural feature), and there is a potential risk that modifying the timing component may accidentally introduce an error in the functional component. In addition, the functional model tends to change very little, as mentioned before; however, the timing model may change a lot during architecture exploration. Hence, it is desirable from a simulator complexity and development point of view to decouple the functional part

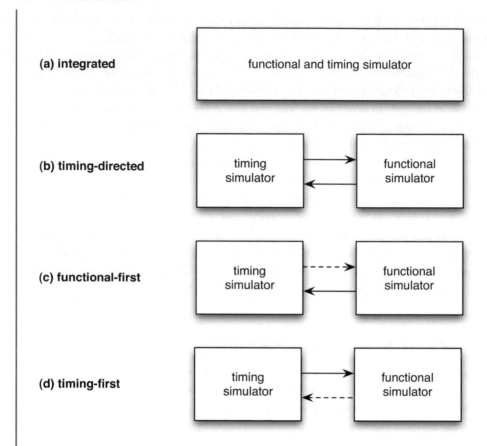

Figure 5.3: Taxonomy of execution-driven simulation.

from the timing part. There are a number of ways of how to do the decoupling, which we discuss now.

Timing-directed simulation.   A timing-directed simulator lets the timing simulator direct the functional simulator to fetch instructions along mispredicted paths and select a particular thread interleaving (Figure 5.3(b)). The Asim simulator [57] is a timing-directed simulator. The functional models keeps track of the architecture state such as register and memory values. The timing model has no notion of values; instead, it gets the effective addresses from the functional model, which it uses to determine cache hits and misses, access the branch predictor, etc. The functional model can be viewed of as a set of function calls that the timing model calls to perform specific functional tasks at precisely the correct simulated time. The functional model needs to be organized such that it can partially simulate instructions. In particular, the functional simulator needs the ability to decode, execute, perform memory operations, kill, and commit instructions. The timing model then calls the functional model to perform specific tasks at the correct time in the correct order. For example, when simulating the execution of a load instruction on a load unit, the timing model asks the functional model to compute the load's effective address. The address is then sent back to the timing model, which subsequently determines whether this load incurs a cache miss. Only when the cache access returns or when a cache miss returns from memory, according to the timing model, will the functional simulator read the value from memory. This ensures that the load reads the exact same data as the target architecture would. When the load commits in the target architecture, the instruction is also committed in the functional model. The functional model also keeps track of enough internal state so that an instruction can be killed in the functional model when it turns out that the instruction was executed along a mispredicted path.

Functional-first simulation.   In the functional-first simulation model (Figure 5.3(c)), a functional simulator feeds an instruction trace into a timing simulator. This is similar to trace-driven simulation, except that the trace need not to be stored on disk; the trace may be fed from the functional simulator into the timing simulator through a UNIX pipe.

In order to be able to simulate along mispredicted paths and model timing-dependent inter-thread orderings and dependences, the functional model provides the ability to roll back to restore prior state [178]. In particular, when executing a branch, the functional model does not know whether the branch is mispredicted — only the timing simulator knows — and thus, it will execute only correct-path instructions. When a mispredicted branch is detected in the fetch stage of the timing simulator, the functional simulator needs to be redirected to fetch instructions along the mispredicted branch. This requires that the functional simulator rolls back to the state prior to the branch and feeds instructions along the mispredicted path into the timing simulator. When the mispredicted branch is resolved in the timing model, the functional model needs to roll back again, and then start feeding correct-path instructions into the timing model. In other words, the functional model is speculating upon which path the branch will take, i.e., it speculates that the branch will be correctly predicted by the timing model.

As mentioned earlier, inter-thread dependences may depend on timing, i.e., small changes in timing may change the ordering in which threads acquire a lock and thus may change functionality and performance. This also applies to a functional-first simulator: the timing may differ between the functional model and the timing model, and as a result, the ordering in which the functional model acquires a lock is not necessarily the same as the ordering observed in the timing model. This ordering problem basically boils down to whether loads 'read' the same data in the functional and timing models. Functional-first simulation can thus handle this ordering problem by keeping track of the data read in the functional model and the timing model. The simulator lets the functional model run ahead; however, when the timing model detects that the data a load would read in the target architecture differs from the data read in the functional model, it rolls back the functional model and requests the functional model to re-execute the load with the correct data — this is called speculative functional-first simulation [29]. Addressing the ordering problem comes at the cost of keeping track of the target memory state in the timing simulator and comparing functional/timing simulator data values.

Argollo et al. [4] present the COTSon simulation infrastructure which employs AMD's SimNow functional simulator to feed a trace of instructions into a timing simulator. The primary focus for COTSon is to simulate complex benchmarks, e.g., commodity operating systems and multi-tier applications, as well as have the ability to scale out and simulate large core counts. In the interest of managing simulator complexity and achieving high simulation speeds, COTSon does not provide roll-back functionality but instead implements timing feedback, which lets the timing simulator adjust the speed of the functional simulator to reflect the timing estimates.

In summary, the key advantage of functional-first simulation is that it allows the functional simulator to run ahead of the timing simulator and exploit parallelism, i.e., run the functional and timing simulator in parallel. Relative to timing-directed simulation in which the timing model directs the functional model at every instruction and/or cycle, functional-first simulation improves simulator performance (i.e., reduces evaluation time) and reduces the complexity of the overall simulator.

Timing-first simulation.   Timing-first simulation lets the timing model run ahead of the functional model [136], see Figure 5.3. The timing simulator models architecture features (register and memory state), mostly correctly, in addition to microarchitecture state. This allows for accurately (though not perfectly) modeling speculative execution along mispredicted branches as well as the ordering of inter-thread events. When the timing model commits an instruction, i.e., when the instruction becomes non-speculative, the functional model verifies whether the timing simulator has deviated from the functional model. On a deviation, the timing simulator is repaired by the functional simulator. This means that the architecture state be reloaded and microarchitecture state be reset before restarting the timing simulation. In other words, timing-first simulation consists of an almost correctly integrated execution-driven simulator (the timing simulator) which is checked by a functionally correct functional simulator.

A timing-first simulator is easier to develop than a fully integrated simulator because the timing simulator does not need to implement all the instructions. A subset of instructions that is important to performance and covers the dynamically executed instructions well is sufficient. Compared to timing-directed simulation, timing-first simulation requires less features in the functional simulator while requiring more features in the timing simulator.

5.6.2   DEALING WITH NON-DETERMINISM

An important challenge one has to deal with when simulating multi-threaded workloads on execution-driven simulators is non-determinism — Alameldeen and Wood present a comprehensive evaluation on non-determinism [1]. Non-determinism refers to the fact that small timing variations can cause executions that start from the same initial state to follow different execution paths. Non-determinism occurs both on real hardware as well as in simulation. On real hardware, timing variations arise from a variety of sources such as interrupts, I/O, bus contention with direct memory access (DMA), DRAM refreshes, etc. One can also observe non-determinism during simulation when comparing architecture design alternatives. For example, changes in some of the design parameters (e.g., cache size, cache latency, branch predictor configuration, processor width, etc.) can cause non-determinism for a number of reasons.

- The operating system might make different scheduling decisions across different runs. For example, the scheduling quantum might end before an I/O event in one run but not in another.

- Threads may acquire locks in a different order. This may cause the number of cycles and instructions spent executing spin-lock loop instructions to be different across different architectures.

- Threads that run at different relative speeds may incur different cache coherence traffic as well as conflict behavior in shared resources. For example, the conflict behavior in shared multicore caches may be different across different architecture designs; similarly, contention in the interconnection network may be different.

Non-determinism severely complicates comparing design alternatives during architecture exploration. The timing differences may lead the simulated workload to take different execution paths with different performance characteristics. As a result, it becomes hard to compare design alternatives. If the variation in the execution paths is significant, comparing simulations becomes unreliable because the amount and type of work done differs across different executions. The fundamental question is whether the performance differences observed across the design alternatives are due to differences in the design alternatives or due to differences in the workloads executed. Not addressing this question may lead to incorrect conclusions.

Alameldeen and Wood [1] present a simple experiment that clearly illustrates the need for dealing with non-determinism in simulation. They consider an OLTP workload and observe that performance increases with increasing memory access time, e.g., 84-ns DRAM leads to 7% better performance than 81-ns DRAM. Of course, this does not make sense — no computer architect will conclude that slower memory leads to better performance. Although this conclusion is obvious in this simple experiment, it may be less obvious in more complicated and less intuitive design studies. Hence, we need appropriate counteractions.

There are three potential solutions for how to deal with non-determinism.

Long-running simulations.   One solution is to run the simulation for a long enough period, e.g., simulate minutes of simulated times rather than seconds. Non-determinism is likely to largely vanish for long simulation experiments; however, given that architecture simulation is extremely slow, this is not a viable solution in practice.

Eliminate non-determinism.   A second approach is to eliminate non-determinism. Lepak et al. [125] and Pereira et al. [156] present approaches to provide reproducible behavior of multi-threaded programs when simulating different architecture configurations on execution-driven simulators; whereas Lepak et al. consider full-system simulation, Pereira et al. focus on user-level simulation. These approaches eliminate non-determinism by guaranteeing that the same execution paths be executed: they enforce the same order of shared memory accesses across simulations by introducing artificial stalls; also, interrupts are forced to occur at specific points during the simulation. Both the Lepak et al. and Pereira et al. approaches propose a metric and method for quantifying to what extent the execution was forced to be deterministic. Introducing stalls implies that the same amount of work be done in each simulation; hence, one can compare design alternatives based on a single simulation. The pitfall of enforcing determinism is that it can lead to executions that may never occur in a real system. In other words, for workloads that are susceptible to non-determinism, this method may not be useful. So, as acknowledged by Lepak et al., deterministic simulation must to be used with care.

Note that trace-driven simulation also completely eliminates non-determinism because the instructions in the trace are exactly the same across different systems. However, trace-driven simulation suffers from the same limitation: it cannot appropriately evaluate architectural designs that affect thread interactions.

Statistical methods.   A third approach is to use (classical) statistical methods to draw valid conclusions. Alameldeen and Wood [1] propose to artificially inject small timing variations during simulation. More specifically, they inject small changes in the memory system timing by adding a uniformly distributed random number between 0 and 4 ns to the DRAM access latency. These randomly injected perturbations create a range of possible executions starting from the same initial condition — note that the simulator is deterministic and will always produce the same timing, hence the need for introducing random perturbations. They then run the simulation multiple times and compute the mean across these runs along with its confidence interval.

An obvious drawback of this approach is that it requires multiple simulation runs, which prolongs total simulation time. This makes this approach more time-consuming compared to the approaches that eliminate non-determinism. However, this is the best one can do to obtain reliable performance numbers through simulation. Moreover, multiple (small) simulation runs is likely to be more time-efficient than one very long-running simulation.

5.7   MODULAR SIMULATION INFRASTRUCTURE

Cycle-accurate simulators are extremely complex pieces of software, on a par with the microarchitectures they are modeling. Like in any other big software project, making sure that the software is well structured is crucial in order to keep the development process manageable. Modularity and reusability are two of the key goals in order to improve manageability and speed of model development. Modularity refers to breaking down the performance modeling problem into smaller pieces that can be modeled separately. Reusability refers to reusing individual pieces in different contexts. Modularity and reusability offer many advantages. It increases modeling productivity as well as fidelity in the performance models (because the individual pieces may have been used and validated before in a different contexts); it allows for sharing individual components across projects, even across products and generations in an industrial environment; it facilitates architectural experimentation (i.e., one

Table 5.2: Comparing different simulators in terms of speed, the architectures and microarchitectures they support, and whether they support full-system simulation [30].

Industry simulators				
Simulator	ISA	μarch	speed	OS
Intel	x86-64	Core 2	1-10 KHz	Yes
AMD	x86-64	Opteron	1-10 KHz	Yes
IBM	Power	Power5	200 KIPS	Yes
Freescale	PowerPC	e500	80 KIPS	No
Academic simulators				
PTLSim [199]	x86-64	AMD Athlon	270 KIPS	Yes
Sim-outorder [20]	Alpha	Alpha 21264	740 KIPS	No
GEMS [133]	Sparc	generic	69 KIPS	Yes

can easily exchange components while leaving the rest of the performance model unchanged). All of these benefits lead to a shorter overall development time.

Several simulation infrastructures implement the modularity principle, see for example Asim [57] by Digital/Compaq/Intel, Liberty [183] at Princeton University, MicroLib [159] at INRIA, UNISIM [6] at INRIA/Princeton, and M5 [16] at the University of Michigan. A modular simulation infrastructure typically provides a simulator infrastructure for creating many performance models rather than having a single performance model. In particular, Asim [57] considers modules, which are the basic software components. A module represents either a physical component of the target design (e.g., a cache, branch predictor, etc.) or a hardware algorithm's operation (e.g., cache replacement policy). Each module provides a well-defined interface, which enables module reuse. Developers can contribute new modules to the simulation infrastructure as long as they implement the module interface, e.g., a branch predictor should implement the three methods for a branch predictor: get a branch prediction, update the branch predictor and handle a mispredicted branch. Asim comes with the Architect's Workbench [58] which allows for assembling a performance model by selecting and connecting modules.

5.8   NEED FOR SIMULATION ACCELERATION

Cycle-accurate simulation is extremely slow, which is a key concern today in architecture research and development. The fundamental reason is that the microarchitectures they are modeling are extremely complex. Today's processors consist of hundreds of millions or even billions of transistors, and they implement complex functionality such as memory hierarchies, speculative execution, out-of-order execution, prefetching, etc. Moreover, the trend towards multicore processing has further exacerbated this problem because multiple cores now need to be simulated as well as their interactions in the shared resources. This huge number of transistors leads to a very large and complex design space that needs to be explored during the design cycle of new microarchitectures.

Although intuition and analytical modeling can help guide the design process, eventually architects have to rely on detailed cycle-accurate simulation in order to make correct design decisions in this complex design space. The complexity of microarchitectures obviously reflects itself in the complexity and speed of the performance models. Cycle-accurate simulation models are extremely slow. Chiou et al. [30] give an overview of the simulation speeds that are typical for today's simulators, see also Table 5.2. The speed of academic simulators ranges between 69 KIPS and 740 KIPS, and they are typically faster than the simulators used in industry which operate in the 1 KHz to 200 KIPS speed range. In other words, simulating only one second of real time (of the target system) may lead to multiple hours or even days of simulation time, even on today's fastest simulators running on today's fastest machines. And this is to simulate a single design point only. Architects typically run many simulations in order to get insight in the design space. Given that the design space is huge, the number of simulations that need to be run is potentially very large, which may make design space exploration quickly become infeasible.

To make things even worse, the benchmarks that are being simulated grow in complexity as well. Given that processors are becoming more and more powerful, the benchmarks need to grow more complex. In the past, before the multicore era, while single-threaded performance was increasing exponentially, the benchmarks needed to execute more instructions and access more data in order to stress contemporary and future processors in a meaningful way. For example, the SPEC CPU benchmarks have grow substantially in complexity: the dynamic instruction count has increased from an average 2.5 billion dynamically executed instructions per benchmark in CPU89 to 230 billion instructions in CPU2000 [99], and an average 2,500 billion instructions per benchmark in CPU2006 [160]. Now, in the multicore era, benchmarks will need to be increasingly multi-threaded in order to stress multicore processors under design.

The slow speed of cycle-accurate simulation is a well-known and long-standing problem, and researchers have proposed various solutions for addressing this important issue. Because of its importance, the rest of this book is devoted to techniques that increase simulation speed. Sampled simulation is covered in Chapter 6 and probably is the most widely used simulation acceleration technique and reduces simulation time by simulating only (a) small snippet(s) from a much longer running benchmark. Statistical simulation, which we revisit in Chapter 7, takes a different approach by generating small synthetic workloads that are representative for long running benchmarks, while at the same time reducing the complexity of the simulator — the purpose for statistical simulation is merely to serve as a fast design space exploration technique that is complementary to detailed cycle-accurate simulation. Finally, in Chapter 8, we describe three approaches that leverage parallelism to speed up simulation: (i) distribute simulation runs across a parallel machine, (ii) parallelize the simulator to benefit from parallelism in the host machine, and (iii) exploit fine-grain parallelism by mapping (parts of) a simulator on reconfigurable hardware (i.e., FPGAs).

CHAPTER 6

Sampled Simulation

The most prevalent method for speeding up simulation is sampled simulation. The idea of sampled simulation is to simulate the execution of only a small fraction of a benchmark's dynamic instruction stream, rather than its entire stream. By simulating only a small fraction, dramatic simulation speedups can be achieved.

Figure 6.1 illustrates the concept of sampled simulation. Sampled simulation simulates one or more so called *sampling units* selected at various places from the benchmark execution. The collection of sampling units is called the *sample*. We refer to the *pre-sampling units* as the parts between two sampling units. Sampled simulation only reports performance metrics of interest for the instructions in the sampling units and discards the instructions in the pre-sampling units. And this is where the dramatic performance improvement comes from: only the sampling units, which is a small fraction of the total dynamic instruction count, are simulated in a cycle-accurate manner.

There are three major challenges for sampled simulation to be accurate and fast:

1. **What sampling units to select?**
   The first challenge is to select a sample that is representative for the entire simulation. Selecting sampling units from the initialization phase of a program execution is unlikely to be representative. This is a manifestation of the more general observation that a program typically goes through various phases of execution. Sampling should reflect this. In other words, a sample needs to be chosen such that all major phases of a program execution are represented.

2. **How to initialize a sampling unit's architecture starting image?**
   The sampling unit's *Architecture Starting Image (ASI)* is the architecture state, i.e., register and memory state, needed to correctly functionally simulate the sampling unit's execution. This is a non-issue for trace-driven simulation because the pre-sampling unit instructions can simply be discarded from the trace, i.e., need not to be stored on disk. For execution-driven simulation, on the other hand, getting the correct ASI in an efficient manner is challenging. It is important to establish the architecture state (registers and memory) as fast as possible so that the sampled simulation can quickly jump from one sampling unit to the next.

3. **How to accurately estimate a sampling unit's microarchitecture starting image?**
   The sampling unit's *Microarchitecture Starting Image (MSI)* is the microarchitecture state (content of caches, branch predictor, processor core structures, etc.) at the beginning of the sampling unit. Obviously, the MSI should be close to the microarchitecture state should the whole dynamic instruction stream prior to the sampling unit be simulated in detail. Unfortunately,

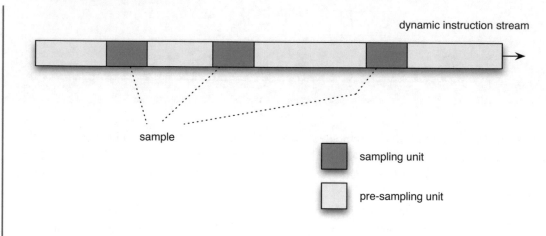

Figure 6.1: General concept of sampled simulation.

during sampled simulation, the sampling unit's MSI is unknown. This is well known in the literature as the *cold-start* problem.

Note the subtle but important difference between getting the architecture state *correct* and getting the microarchitecture state *accurate*. Getting the architecture state correct is absolutely required in order to enable a functionally correct execution of the sampling unit. Getting the microarchitecture state as accurate as possible compared to the case where the entire dynamic instruction stream would have been executed up to the sampling unit is desirable if one wants accurate performance estimates.

The following subsections describe each of these three challenges in more detail.

6.1 WHAT SAMPLING UNITS TO SELECT?

There are basically two major ways for selecting sampling units, statistical sampling and targeted sampling.

6.1.1 STATISTICAL SAMPLING

Statistical sampling can select the sampling units either randomly or periodically. Random sampling selects sampling units randomly from the entire instruction stream in an attempt to provide a unbiased and representative sample, see Figure 6.2(a). Laha et al. [113] pick randomly chosen sampling units for evaluating cache performance. Conte et al. [36] generalized this concept to sampled processor simulation.

The SMARTS (Sampling Microarchitecture Simulation) approach by Wunderlich et al. [193; 194] proposes systematic sampling, also called periodic sampling, which selects sampling units

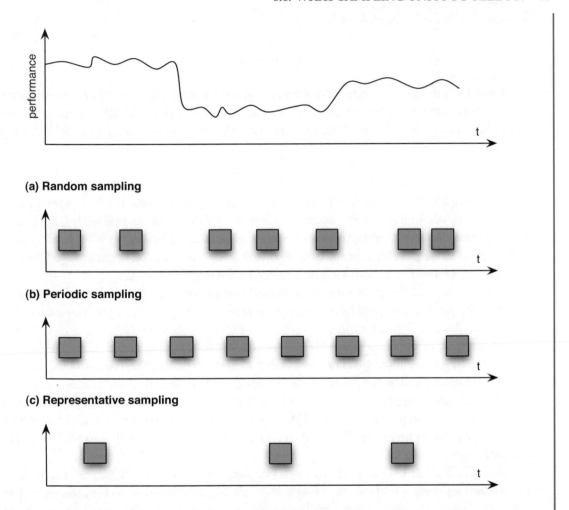

(a) Random sampling

(b) Periodic sampling

(c) Representative sampling

Figure 6.2: Three ways for selecting sampling units: (a) random sampling, (b) periodic sampling, and (c) representative sampling.

periodically across the entire program execution, see also Figure 6.2(b): the pre-sampling unit size is fixed, as opposed to random sampling, e.g., a sampling unit of 10,000 instructions is selected every 1 million instructions.

The key advantage of statistical sampling is that it builds on statistics theory, and allows for computing confidence bounds on the performance estimates through the central limit theorem [126]. Assume we have a performance metric of interest (e.g., CPI) for n sampling units: x_i, $1 \leq i \leq n$.

The mean of these measurements $\bar{x}$ is computed as

$$\bar{x} = \frac{\sum_{i=1}^{n} x_i}{n}.$$

The central limit theory then states that, for large values of n (typically $n \geq 30$), $\bar{x}$ is approximately Gaussian distributed provided that the samples x_i are (i) independent,[1] and (ii) come from the same population with a finite variance.[2] Statistics then states that we can compute the confidence interval $[c_1, c_2]$ for the mean as

$$[\bar{x} - z_{1-\alpha/2}\frac{s}{\sqrt{n}}; \bar{x} + z_{1-\alpha/2}\frac{s}{\sqrt{n}}],$$

with s the sample's standard deviation. The value $z_{1-\alpha/2}$ is typically obtained from a precomputed table; $z_{1-\alpha/2}$ equals 1.96 for a 95% confidence interval. A 95% *confidence interval* $[c_1, c_2]$ basically means that the probability for the true mean μ to lie between c_1 and c_2 equals 95%. In other words, a confidence interval gives the user some confidence that the true mean (e.g., the average CPI across the entire program execution) can be approximated by the sample mean $\bar{x}$.

SMARTS [193; 194] goes one step further and leverages the above statistics to determine how many sampling units are required to achieve a desired confidence interval at a given confidence level. In particular, the user first determines a particular confidence interval size (e.g., a 95% confidence interval within 3% of the sample mean). The benchmark is then simulated and n sampling units are collected, n being some initial guess for the number of sampling units. The mean and its confidence interval is computed for the sample, and, if it satisfies the above 3% rule, this estimate is considered to be good. If not, more sampling units ($> n$) must be collected, and the mean and its confidence interval must be recomputed for each collected sample until the accuracy threshold is satisfied. This strategy yields bounded confidence interval sizes at the cost of requiring multiple (sampled) simulation runs.

SMARTS [193; 194] uses a fairly small sampling unit size of 1,000 instructions for SPEC CPU workloads; Flexus [190] uses sampling units of a few 10,000 instructions for full-system server workloads. The reason for choosing a small sampling unit size is to minimize measurement (reduce number of instructions simulated in detail) while keeping into account measurement practicality (i.e., measure IPC or CPI over a long enough time period) and bias (i.e., make sure microarchitecture state is warmed up, as we will discuss in Section 6.3). The use of small sampling units implies that we need lots of them, typically on the order of 1000 sampling units. The large number of sampling units implies in its turn that statistical simulation becomes embarrassingly parallel, i.e., one can distribute the sampling units across a cluster of machines, as we will discuss in Section 8.1. In addition, it allows for throttling simulation turnaround time on-the-fly based on a desired error and confidence.

---

[1] One could argue whether the condition of independent measurements is met because the sampling units are selected from a single program execution, and thus these measurements are not independent. This is even more true for periodic sampling because the measurements are selected at fixed intervals.

[2] Note that the central limit theory does not impose a particular distribution for the population from which the sample is taken. The population may not be Gaussian distributed (which is most likely to be the case for computer programs), yet the sample mean is Gaussian distributed.

The potential pitfall of systematic sampling compared to random sampling is that the sampling units may give a skewed view in case the periodicity present in the program execution under measurement equals the sampling periodicity or its higher harmonics. For populations with low homogeneity though, periodic sampling is a good approximation of random sampling. Wunderlich et al. [193; 194] showed this to be the case for their workloads. This also agrees with the intuition that the workloads do not have sufficiently regular cyclic behavior at the periodicity relevant to sampled simulation (tens of millions of instructions).

6.1.2 TARGETED SAMPLING

Targeted sampling contrasts with statistical sampling in that it first analyzes the program's execution to pick a sampling unit for each unique behavior in the program's execution, see also Figure 6.2(c): targeted sampling selects a single sampling unit from each program phase and then weighs each sampling unit to provide an overall performance number. The key advantage of targeted sampling relative to statistical sampling is that it may lead to potentially fewer sampling units and thus a shorter overall simulation time as well as an overall simpler setup. The reason is that it leverages program analysis and intelligently picks sampling units. The major limitation for targeted sampling is its inability to provide a confidence bound on the performance estimates, unlike statistical sampling. A number of approaches have been proposed along this line, which we briefly revisit here.

Skadron et al. [172] select a single sampling unit of 50 million instructions for their microarchitectural simulations. To this end, they first measure branch misprediction rates, data cache miss rates and instruction cache miss rates for each interval of 1 million instructions. By plotting these measurements as a function of the number of instructions simulated, they observe the time-varying program execution behavior, i.e., they can identify the initialization phase and/or periodic behavior in a program execution. Based on these plots, they manually select a contiguous sampling unit of 50 million instructions. Obviously, this sampling unit is chosen after the initialization phase. The validation of the 50 million instruction sampling unit is done by comparing the performance characteristics (obtained through detailed architectural simulations) of this sampling unit to 250 million instruction sampling units. The selection and validation of a sampling unit is done manually. The potential pitfall of this approach is that although the sampling unit is representative for a larger instruction sequence for this particular microarchitecture, it may not be representative on other microarchitectures. The reason is that the similarity analysis was done for a particular microarchitecture. In other words, phase analysis and picking the sampling unit is done based on microarchitecture-dependent characteristics and performance metrics only.

Lafage and Seznec [112] use cluster analysis to detect and select sampling units that exhibit similar behavior. In their approach, they first measure two microarchitecture-independent metrics for each instruction interval of one million instructions. These metrics quantify the temporal and spatial behavior of the data reference stream in each instruction interval. Subsequently, they perform cluster analysis and group intervals that exhibit similar temporal and spatial behavior into so called clusters. For each cluster, the sampling unit that is closest to the center of the cluster is chosen as the representative sampling unit for that cluster. The performance characteristics per representative sampling unit are then weighted with the number of sampling units it represents, i.e., with the number of intervals grouped in the cluster the sampling unit represents. The microarchitecture-independent metrics proposed by Lafage and Seznec are limited to quantifying data stream locality only.

SimPoint [171] is the most well known targeted sampling approach, see Figure 6.3 for an overview of the approach. SimPoint breaks a program's execution into intervals, and for each interval, it creates a code signature (step 1 in Figure 6.3). (An interval is a sequence of dynamically executed instructions.) The code signature is a so called Basic Block Vector (BBV) [170], which counts the number of times each basic block is executed in the interval, weighted with the number of instructions per basic block. Because a program may have a large static code footprint, and thus may touch (i.e., execute at least once) a large number of basic blocks, the BBVs may be large. SimPoint therefore reduces the dimensionality of the BBVs through random projection (step 2 in Figure 6.3), i.e., the dimensionality of the BBVs is reduced to a 15-dimensional vector, in order to increase the effectiveness of the next step. SimPoint then performs clustering, which aims at finding the groups of intervals that have similar BBV behavior (step 3 in Figure 6.3). Intervals from different parts of the program execution may be grouped into a single cluster. A cluster is also referred to as a phase. The key idea behind SimPoint is that intervals that execute similar code, i.e., have similar BBVs, have similar architecture behavior (e.g., cache behavior, branch behavior, IPC performance, power consumption, etc.). And this has been shown to be the case, see [116]. Therefore, only one interval from each phase needs to be simulated in order to recreate a complete picture of the program's execution. They then choose a representative sampling unit from each phase and perform detailed simulation on that interval (step 4 in Figure 6.3). Taken together, these sampling units (along with their respective weights) can represent the complete execution of a program. In SimPoint terminology, a sampling unit is called a *simulation point*. Each simulation point is an interval of on the order of millions, or tens to hundreds of millions of instructions. Note that the simulation points were selected by examining only a profile of the code executed by a program. In other words, the profile is microarchitecture-independent. This suggests that the selected simulation points can be used across microarchitectures (which people have done successfully); however, there may be a potential pitfall in that different microarchitecture features may lead to a different performance impact for the simulation points compared to the parts of the execution not selected by SimPoint, and this performance impact is potentially unbounded — statistical sampling, on the other hand, bounds the error across microarchitectures.

The SimPoint group extended SimPoint in a number of ways. They proposed techniques for finding simulation points early in the dynamic instruction stream in order to reduce the time needed to functionally simulate to get to these simulation points [157]. They considered alternative program characteristics to BBVs to find representative sampling units, such as loops and method calls [115; 117]. They correlate a program's phase behavior to its control flow behavior, and by doing so, they identify cross binary simulation points so that simulation points can be used by

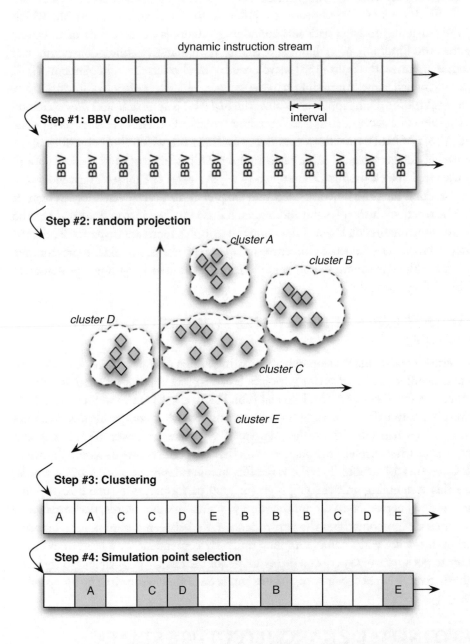

Figure 6.3: Overview of the SimPoint approach.

architects and compiler builders when studying ISA extensions and evaluating compiler and software optimizations [158]. SimPoint v3.0 [77] improves the efficiency of the clustering algorithm, which enables applying the clustering to large data sets containing hundreds of thousands of intervals; the end result is that the SimPoint procedure for selecting representative simulation points can be applied in a couple minutes. PinPoint [153], developed by Intel researchers, implements BBV collection as a Pin tool, which allows for finding representative simulation points in x86 workloads. Yi et al. [195] compared the SimPoint approach against the SMARTS approach, and they conclude that SMARTS is slightly more accurate than SimPoint, however, SimPoint has a better speed versus accuracy trade-off. Also, SMARTS provides a confidence on the error, which SimPoint does not.

Other approaches in the category of targeted sampling include early work by Dubey and Nair [42], which uses basic block profiles; Lauterbach [118] evaluates the representativeness of a sample using its instruction mix, the function execution frequency and cache statistics; Iyengar et al. [91] consider an instruction's history (cache and branch behavior of the instructions prior to the instruction in the dynamic instruction stream) to quantify a sample's representativeness. Eeckhout et al. [51] consider a broad set of microarchitecture-independent metrics to find representative sampling units across multiple programs; the other approaches are limited to finding representative sampling unit within a single program.

6.1.3 COMPARING DESIGN ALTERNATIVES THROUGH SAMPLED SIMULATION

In computer architecture research and development, comparing design alternatives, i.e., determining the relative performance difference between design points, is often more important than determining absolute performance in a single design point. Luo and John [130], and Ekman and Stenström [54] make the interesting observation that fewer sampling units are needed when comparing design points than when evaluating performance in a single design point. They make this observation based on a matched-pair comparison, which exploits the phenomenon that the performance difference between two designs tends to be (much) smaller than the variability across sampling units. In other words, it is likely the case that a sampling unit yielding high (or low) performance for one design point will also yield high (or low) performance for another design point. In other words, performance is usually correlated across design points, and the performance ratio between design points does not change as much as performance varies across sampling units. By consequence, we need to evaluate fewer sampling units to get an accurate performance estimate for the relative performance difference between design alternatives. Matched-pair comparisons can reduce simulation time by an order of magnitude [54; 130].

6.2 HOW TO INITIALIZE ARCHITECTURE STATE?

The second issue to deal with in sampled processor simulation is how to accurately provide a sampling unit's architecture starting image (ASI). The ASI is the architecture state (register and memory content) needed to functionally simulate the sampling unit's execution to achieve the correct output for that sampling unit. This means that the register and memory state needs to be established at the beginning of the sampling unit just as if all instructions prior to the sampling unit would have been executed. The two approaches for constructing the ASI are fast-forwarding and checkpointing, which we discuss in more detail.

6.2.1 FAST-FORWARDING

The principle of fast-forwarding between sampling units is illustrated in Figure 6.4(a). Starting from either the beginning of the program or the prior sampling unit, fast-forwarding constructs the architecture starting image through functional simulation. When it reaches the beginning of the next sampling unit, the simulator switches to detailed execution-driven simulation. When detailed simulation reaches the end of the sampling unit, the simulator switches back to functional simulation to get to the next sampling unit, or in case the last sampling unit is executed, the simulator quits.

The main advantage is that it is relatively straightforward to implement in an execution-driven simulator — an execution-driven simulator comes with a functional simulator, and switching between functional simulation and detailed execution-driven simulation is not that hard to implement. The disadvantage is that fast-forwarding can be time-consuming for sampling units that are located deep in the dynamic instruction stream. In addition, it also serializes the simulation of all of the sampling units, i.e., one needs to simulate all prior sampling units and fast-forward between the sampling units in order to construct the ASI for the next sampling unit. Because fast-forwarding can be fairly time-consuming, researchers have proposed various techniques to speed up fast-forwarding.

Szwed et al. [179] propose to fast-forward between sampling units through native hardware execution, called *direct execution*, rather than through functional simulation, see also Figure 6.4(b). Because native hardware execution is much faster than functional simulation, substantial speedups can be achieved. Direct execution is employed to quickly go from one sampling unit to the next. When the next sampling unit is reached, checkpointing is used to communicate the architecture state from the real hardware to the simulator. Detailed execution-driven simulation of the sampling unit is done starting from this checkpoint. When the end of the sampling unit is reached, the simulator switches back to native hardware execution to quickly reach the next sampling unit. Many ways to incorporate direct hardware execution into simulators for speeding up simulation and emulation systems have been proposed, see for example [43; 70; 109; 163; 168].

One requirement for fast-forwarding through direct execution is that the simulation needs to be done on a host machine with the same instruction-set architecture (ISA) as the target machine. Fast-forwarding on a host machine with a different ISA than the target machine cannot be sped up through direct execution. This is a serious concern for studies that explore ISA extensions, let alone an entirely novel ISA. This would imply that such studies would need to fall back to relatively slow functional simulation. One possibility to overcome this limitation is to employ techniques from dynamic binary translation methods such as just-in-time (JIT) compilation and caching of translated code, as is done in Embra [191]. A limitation with dynamic binary translation though is that it makes the simulator less portable to host machines with different ISAs. An alternative

(a) Fast-forwarding through functional simulation

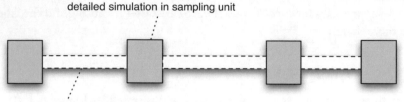

detailed simulation in sampling unit

functional simulation between sampling units

(b) Fast-forwarding through direct execution

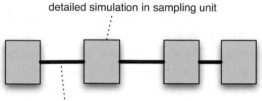

detailed simulation in sampling unit

native hardware execution between sampling units

(c) Checkpointing

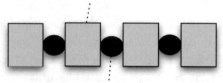

detailed simulation in sampling unit

loading the checkpoint from disk

Figure 6.4: Three approaches to initialize the architecture starting image: (a) fast-forwarding through functional simulation, (b) fast-forwarding through direct execution, and (c) checkpointing.

approach is to resort to so called compiled instruction-set simulation as proposed by [21; 147; 164]. The idea of compiled instruction-set simulation is to translate each instruction in the benchmark by C code that decodes the instruction. Compiling the C code yields a functional simulator. Given that the generated functional simulator is written in C, it is easily portable across platforms. (We already discussed these approaches in Section 5.2.1.)

6.2.2   CHECKPOINTING

Checkpointing takes a different approach and stores the ASI before a sampling unit. Taking a checkpoint is similar to storing a core dump of a program so that it can be replayed at that point in execution. A checkpoint stores the register contents and the memory state prior to a sampling unit. During sampled simulation, getting the architecture starting image initialized is just a matter of loading the checkpoint from disk and updating the register and memory state in the simulator, see Figure 6.4(c). The advantage of checkpointing is that it allows for parallel simulation, in contrast to fast-forwarding, i.e., checkpoints are independent of each other and enables simulating multiple sampling units in parallel.

There is one major disadvantage to checkpointing compared to fast-forwarding and direct execution, namely, large checkpoint files need to be stored on disk. Van Biesbrouck et al. [184] report checkpoint files up to 28 GB for a single benchmark. Using many sampling units could be prohibitively costly in terms of disk space. In addition, the large checkpoint file size also affects total simulation time due to loading the checkpoint file from disk when starting the simulation of a sampling unit and transferring over a network during parallel simulation.

Reduced checkpointing addresses the large checkpoint concern by limiting the amount of information stored in the checkpoint. The main idea behind reduced checkpointing is to only store the registers along with the memory words that are read in the sampling unit — a naive checkpointing approach would store the entire memory state. The Touched Memory Image (TMI) approach [184] and the live-points approach in TurboSMARTS [189] implement this principle. The checkpoint only stores the chunks of memory that are read during the sampling unit. This is a substantial optimization compared to full checkpointing which stores the entire memory state for each sampling unit. An additional optimization is to store only the chunks of memory that are read before they are written — there is no need to store a chunk of memory in the checkpoint in case that chunk of memory is written prior to being read in the sampling unit. At simulation time, prior to simulating the given sampling unit, the checkpoint is loaded from disk and the chunks of memory in the checkpoint are written to their corresponding memory addresses. This guarantees a correct ASI when starting the simulation of the sampling unit. A small file size is further achieved by using a sparse image representation, so regions of memory that consist of consecutive zeros are not stored in the checkpoint.

Van Biesbrouck et al. [184] and Wenisch et al. [189] provide a comprehensive evaluation of the impact of reduced ASI checkpointing on simulation accuracy, storage requirements, and simulation time. These studies conclude that the impact on error is marginal (less than 0.2%) — the reason for the inaccuracy due to ASI checkpointing is that the data values for loads along mispredicted paths may be incorrect. Reduced ASI checkpointing reduces storage requirements by two orders of magnitude compared to full ASI checkpointing. For example, for SimPoint using one-million instruction sampling units, an average (compressed) full ASI checkpoint takes 49.3 MB whereas a reduced ASI checkpoint takes only 365 KB. Finally, reduced ASI checkpointing reduces the simulation time by an order of magnitude (20×) compared to fast-forwarding and by a factor 4× compared to full checkpointing.

Ringenberg and Mudge [165] present intrinsic checkpointing which basically stores the checkpoint in the binary itself. In other words, intrinsic checkpointing brings the ASI up to state by providing fix-up checkpointing code consisting of store instructions to put the correct data values in memory — again, only memory locations that are read in the sampling unit need to be updated; it also executes instructions to put the correct data values in registers. Intrinsic checkpointing has the limitation that it requires binary modification for including the checkpoint code in the benchmark binary. On the other hand, it does not require modifying the simulator, and it even allows for running sampling units on real hardware. Note though that the checkpoint code may skew the performance metrics somewhat; this can be mitigated by considering large sampling units.

6.3   HOW TO INITIALIZE MICROARCHITECTURE STATE?

The third issue in sampled simulation is to establish an as accurate as possible microarchitecture starting image (MSI), i.e., cache state, predictor state, processor core state, etc., for the sampling unit to be simulated. The MSI for the sampling unit should be as accurate as possible compared to the MSI that would have been obtained through detailed simulation of all the instructions prior to the sampling unit. An inaccurate MSI introduces error and will sacrifice the error bound and confidence in the estimates (for statistical sampling).

The following subsections describe MSI approaches related to cache structures, branch predictors, and processor core structures such as the reorder buffer, issue queues, store buffers, functional units, etc.

6.3.1   CACHE STATE WARMUP

Caches are probably the most critical aspect of the MSI because caches can be large (up to several MBs) and can introduce a long history. In this section, we use the term 'cache' to collectively refer to a cache, a Translation Lookaside Buffers (TLB) and a Branch Target Buffers (BTB) because all of these structures have a cache-like structure.

A number of cache state warmup strategies have been proposed over the past 15 years. We now discuss only a selection.

No warmup.   The *cold* or *no warmup* scheme [38; 39; 106] assumes an empty cache at the beginning of each sampling unit. Obviously, this scheme will overestimate the cache miss rate. However, the bias can be small for large sampling unit sizes. Intel's PinPoint approach [153], for example, considers a fairly large sampling unit size, namely 250 million instructions, and does not employ any warmup approach because the bias due to an inaccurate MSI is small.

Continuous warmup.   Continuous warmup, as the name says, continuously keeps the cache state warm between sampling units. This means that the functional simulation between sampling units needs to be augmented to also access the caches. This is a very accurate approach but increases the time spent between sampling units. This approach is implemented in SMARTS [193; 194]: the tiny sampling units of 1,000 instructions used in SMARTS require a very accurate MSI, which is achieved through continuous warmup; this is called *functional warming* in the SMARTS approach.

Stitch.   *Stitch* or *stale state* [106] approximates the microarchitecture state at the beginning of a sampling unit with the hardware state at the end of the previous sampling unit. An important disadvantage of the stitch approach is that it cannot be employed for parallel sampled simulation.

Cache miss rate estimation.   Another approach is to assume an empty cache at the beginning of each sampling unit and to estimate which cold-start misses would have missed if the cache state at the beginning of the sampling unit was known. This is the so called cache miss rate estimator approach [106; 192]. A simple example cache miss estimation approach is hit-on-cold or assume-hit. Hit-on-cold assumes that the first access to a cache line is always a hit. This is an easy-to-implement technique which is fairly accurate for programs with a low cache miss rate.

Self-monitored adaptive (SMA) warmup.   Luo et al. [131] propose a self-monitored adaptive (SMA) cache warmup scheme in which the simulator monitors the warmup process of the caches and decides when the caches are warmed up. This warmup scheme is adaptive to the program being simulated as well as to the cache being simulated — the smaller the application's working set size or the smaller the cache, the shorter the warmup phase. One limitation of SMA is that it is unknown a priori when the caches will be warmed up and when detailed simulation should get started. This may be less of an issue for random statistical sampling (although the sampling units are not selected in a random fashion anymore), but it is a problem for periodic sampling and targeted sampling.

Memory Reference Reuse Latency (MRRL).   Haskins and Skadron [79] propose the MRRL warmup strategy. The memory reference reuse latency is defined as the number of instructions between two consecutive references to the same memory location. The MRRL warmup approach computes the MRRL for each memory reference in the sampling unit, and collects these MRRLs in a distribution. A given percentile, e.g., 99%, then determines when cache warmup should start prior to the sampling unit. The intuition is that a sampling unit with large memory reference reuse latencies also needs a long warmup period.

Boundary Line Reuse Latency (BLRL).   Eeckhout et al. [49] only look at reuse latencies that 'cross' the boundary line between the pre-sampling unit and the sampling unit, hence the name boundary line reuse latency (BLRL). In contrast, MRRL considers all the reuse latencies which may not be an accurate picture for the cache warmup required for the sampling unit. Relative to BLRL, MRRL may result in a warmup period that is either too short to be accurate or too long for the attained level of accuracy.

Checkpointing.   Another approach to the cold-start problem is to *checkpoint* or to store the MSI at the beginning of each sampling unit. Checkpointing yields perfectly warmed up microarchitecture state. On the flipside, it is specific to a particular microarchitecture, and it may require excessive disk space for storing checkpoints for a large number of sampling units and different microarchitectures. Since this is infeasible to do in practice, researchers have proposed more efficient approaches to MSI checkpointing.

One approach is the No-State-Loss (NSL) approach [35; 118]. NSL scans the pre-sampling unit and records the last reference to each unique memory location. This is called the *least recently used (LRU) stream*. For example, the LRU stream of the following reference stream 'ABAACDABA' is 'CDBA'. The LRU stream can be computed by building the LRU stack: it is easily done by pushing an address on top of the stack when it is referenced. NSL yields a perfect warmup for caches with an LRU replacement policy.

Barr and Asanovic [10] extended this approach for reconstructing the cache and directory state during sampled multiprocessor simulation. In order to do so, they keep track of a timestamp per unique memory location that is referenced. In addition, they keep track of whether a memory location is read or written. This information allows them to quickly reconstruct the cache and directory state at the beginning of a sampling unit.

Van Biesbrouck et al. [185] and Wenisch et al. [189] proposed a checkpointing approach in which the largest cache of interest is simulated once for the entire program execution. The SimPoint project refers to this technique as 'memory hierarchy state'; the TurboSMARTS project proposes the term 'live points'. At the beginning of each sampling unit, the cache content is stored on disk as a checkpoint. The content of smaller sized caches can then be derived from the checkpoint. Constructing the content of a cache with a smaller associativity is trivial to from the checkpoint: the most recently accessed cache lines need to be retained per set, see Figure 6.5(a). Reducing the number of sets in the cache is slightly more complicated: the new cache set retains the most recently used cache lines from the merging cache sets — this requires keeping track of access times to cache lines during checkpoint construction, see Figure 6.5(b).

6.3.2 PREDICTOR WARMUP

Compared to the amount of work done on cache state warmup, little work has been done on predictor warmup — it is an overlooked problem. In particular, accurate branch predictor warmup is required for accurate sampled simulation, even for fairly large sampling units of 1 to 10 million instructions [107]. There is no reason to believe that this observation made for branch predictors does not generalize to other predictors, such as next cache line predictors, prefetchers, load hit/miss predictors, load/store dependency predictors, etc.

One approach is to employ stitch, i.e., the sampling unit's MSI is assumed to be the same as the state at the end of the prior sampling unit. Another approach is to consider a fixed-length warmup, e.g., start warming the branch predictor at for example 1 million instructions prior to the sampling unit, as proposed by Conte et al. [36]. Barr and Asanovic [9] propose warming the branch predictor using all the instructions prior to the sampling unit. In order not to have to store huge branch trace files to be stored on disk, they propose branch trace compression. Kluyskens and Eeckhout [107] propose Branch History Matching (BHM) which builds on a similar principle as MRRL and BLRL. BHM considers the reuse latency between dynamic branch instances that have share the same (or at least a very similar) branch history. Once the reuse latency distribution is computed, it is determined how long predictor warmup should take prior to the sampling unit.

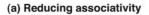

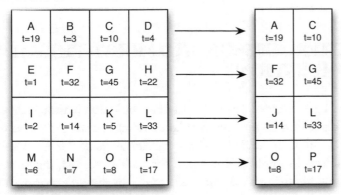

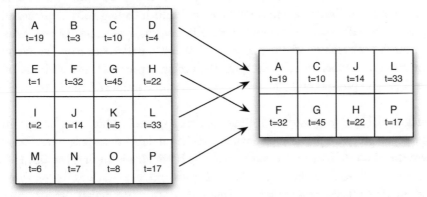

Figure 6.5: Constructing the content of a smaller sized cache from a checkpoint, when (a) reducing associativity and (b) reducing the number of sets. Each cache line in the checkpoint is tagged with a timestamp that represents the latest access to the cacheline.

6.3.3 PROCESSOR CORE STATE

So far, we discussed MSI techniques for cache and branch predictor structures. The processor core consists of a reorder buffer, issue queues, store buffers, functional units, etc., which may also need to be warmed up. This is not a major concern for large sampling units because events in the processor core do not incur an as long history as in the cache hierarchy and branch predictors. However, for small sampling units, it is crucial to accurately warmup the processor core. Therefore, SMARTS [193; 194] considers small sampling units of 1,000 instructions and proposes fixed-length warming of the processor core of 2,000 to 4,000 instructions prior to each sampling unit — warmup length can be fixed because of the bounded history of the processor core as opposed to the unbounded history for caches, TLBs and predictors.

6.4   SAMPLED MULTIPROCESSOR AND MULTI-THREADED PROCESSOR SIMULATION

Whereas sampled simulation for single-threaded processors can be considered mature technology, and significant progress has been made towards sampled simulation of multi-threaded server workloads (see the Flexus project [190]), sampling multi-threaded workloads in general remains an open problem. One key problem to address in sampled simulation for multi-threaded, multicore and multiprocessor architectures running multi-threaded workloads relates to resource sharing. When two or more programs or threads share a processor's resource such as a shared L2 cache or interconnection network — as is the case in many contemporary multi-core processors — or even issue queues and functional units — as is the case in Simultaneous Multithreading (SMT) processors — the performance of both threads becomes entangled. In other words, co-executing programs and threads affect each other's performance. And, changing a hardware parameter may change which parts of the program execute together, thereby changing their relative progress rates and thus overall system performance. This complicates both the selection of sampling units and the initialization of the ASI and MSI.

Van Biesbrouck et al. [188] propose the co-phase matrix approach, which models the impact of resource sharing on per-program performance when running independent programs (i.e., from a multi-program workload) on multithreaded hardware. The basic idea is to first use SimPoint to identify the program phases for each of the co-executing threads and keep track of the performance data of previously executed co-phases in a so called *co-phase matrix*; whenever a co-phase gets executed again, the performance data is easily picked from the co-phase matrix. By doing so, each unique co-phase gets simulated only once, which greatly reduces the overall simulation time. The co-phase matrix is an accurate and fast approach for estimating multithreaded processor performance both when the co-executing threads start at a given starting point as well as when multiple starting points are considered for the co-executing threads, see [186]. The multiple starting points approach provides a much more representative overall performance estimate than a single starting point. Whereas the original co-phase matrix work focuses on two or four independent programs co-executing on a multithreaded processor, the most recent work by Van Biesbrouck et al. [187] studies how to select a limited number of representative co-phase combinations across multiple benchmarks within a benchmark suite.

As mentioned earlier, Barr and Asanovic [10] propose an MSI checkpointing technique to reconstruct the directory state at the beginning of a sampling unit in a multiprocessor system. State reconstruction for a multiprocessor is harder than for a single-core processor because the state is a function of the relative speeds of the programs or threads running on the different cores, which is hard to estimate without detailed simulation.

Whereas estimating performance of a multi-threaded workload through sampled simulation is a complex problem, estimating overall system performance for a set if independent threads or programs is much simpler. Ekman and Stenström [54] observed that the variability on overall system throughput is smaller than the per-thread performance variability when running multiple independent threads concurrently. The intuitive explanation is that there is a smoothing effect of different threads executing high-IPC and low-IPC phases simultaneously. As such, if one is interested in overall system throughput, a relatively small sample size will be enough to obtain accurate average performance estimates and performance bounds; Ekman and Stenström experimentally verify that a factor N fewer sampling units are needed when simulating a system with N cores compared to single-core simulation. This is true only if the experimenter is interested in average system throughput only. If the experimenter is interested in the performance for individual programs, he/she will need to collect more sampling units. In addition, this smoothing effect assumes that the various threads are independent. This is the case, for example, in commercial transaction-based workloads where transactions, queries and requests arrive randomly, as described by Wenisch et al. [190].

CHAPTER 7

Statistical Simulation

Statistical modeling has a long history. Researchers typically employ statistical modeling to generate synthetic workloads that serve as proxies for realistic workloads that are hard to capture. For example, collecting traces of wide area networks (WAN) or even local area networks (LAN) (e.g., a cluster of machines) is non-trivial, and it requires a large number of disks to store these huge traces. Hence, researchers often resort to synthetic workloads (that are easy to generate) to exhibit the same characteristics (in a statistical sense) as the real network traffic. As another example, researchers studying commercial server systems may employ statistical workload generators to generate load for the server systems under study. Likewise, researchers in the area of interconnection networks frequently use synthetic workloads in order to evaluate a network topology and/or router design across a range of network loads.

Synthetic workload generation can also be used to evaluate processor architectures. For example, Kumar and Davidson [111] used synthetically generated workloads to evaluate the performance of the memory subsystem of the IBM 360/91; the motivation for using synthetic workloads is that they enable investigating the performance of a computer system as a function of the workload characteristics. Likewise, Archibald and Baer [3] use synthetically generated address streams to evaluate cache coherence protocols. The paper by Carl and Smith [23] renewed recent interest in synthetic workloads for evaluating modern processors and coined the term 'statistical simulation'. The basic idea of statistical simulation is to collect a number of program characteristics in the form of distributions and then generate a synthetic trace from it that serves as a proxy for the real program. Simulating the synthetic trace then yields a performance estimate for the real program. Because the synthetic trace is much shorter than the real program trace, simulation is much faster. Several research projects explored this idea over the past decade [13; 50; 87; 149; 152].

7.1 METHODOLOGY OVERVIEW

Figure 7.1 illustrates the general framework of statistical simulation, which consists of three major steps. The first step collects a number of execution characteristics for a given workload in the form of distributions, hence the term statistical profiling. The statistical profile typically consists of a set of characteristics that are independent of the microarchitecture (e.g., instruction mix, instruction dependencies, control flow behavior) along with a set of microarchitecture-dependent characteristics (typically locality events such as cache and branch prediction miss rates). Statistical profiling is done only once and is done relatively quickly through specialized simulation, which is much faster (typically more than one order of magnitude) compared to cycle-accurate simulation. The second step

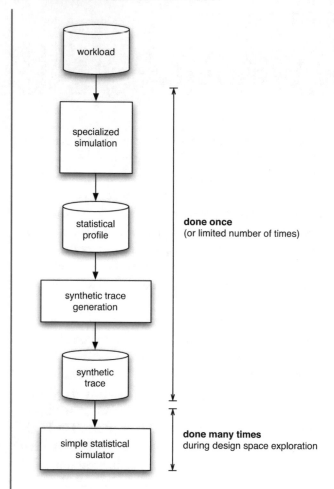

Figure 7.1: Statistical simulation framework.

generates a synthetic trace based from this statistical profile. The characteristics of the trace reflect the properties in the statistical profile and thus the original workload, by construction. Finally, simulating the synthetic trace on a simple trace-driven statistical simulator yields performance numbers. The hope/goal is that, if the statistical profile captures the workload's behavior well and if the synthetic trace generation algorithm is able to translate these characteristics into a synthetic trace, then the performance numbers obtained through statistical simulation should be accurate estimates for the performance numbers of the original workload.

The key idea behind statistical simulation is that capturing a workload's execution behavior in the form of distributions enables generating short synthetic traces that are representative for long-running real-life applications and benchmarks. Several researchers have found this property to hold true: it is possible to generate short synthetic traces with on the order of a few millions of instructions that resemble workloads that run for tens to hundreds of billions of instructions — this implies a simulation speedup of at least four orders of magnitude. Because the synthetic trace is generated based on distributions, its performance characteristics quickly converge, typically after one million (or at most a few millions) of instructions. And this is obviously where the key advantage lies for statistical simulation: it enables predicting performance for long-running workloads using short running synthetic traces. This is likely to speed up processor architecture design space exploration substantially.

7.2   APPLICATIONS

Statistical simulation has a number of potential applications.

Design space exploration.   The most obvious application for statistical simulation is processor design space exploration. Statistical simulation does not aim at replacing detailed cycle-accurate simulation, primarily because it is less accurate — e.g., it does not model cache accesses along mispredicted paths, it simulates an abstract representation of a real workload, etc., as we will describe later. Rather, statistical simulation aims at providing a tool that enables a computer architect to quickly make high-level design decisions, and it quickly steers the design exploration towards a region of interest, which can then be explored through more detailed (and thus slower) simulations. Steering the design process in the right direction early on in the design cycle is likely to reduce the overall design process and time to market. In other words, statistical simulation is to be viewed of as a useful complement to the computer architect's toolbox to quickly make high-level design decisions early in the design cycle.

Workload space exploration.   Statistical simulation can also be used to explore how program characteristics affect performance. In particular, one can explore the workload space by varying the various characteristics in the statistical profile in order to understand how these characteristics relate to performance. The program characteristics that are part of the statistical profile are typically hard to vary using real benchmarks and workloads, if at all possible. Statistical simulation, on the other hand, allows for easily exploring this space. Oskin et al. [152] provide such a case study in which they vary basic block size, cache miss rate, branch misprediction rate, etc. and study its effect on performance. They also study the potential of value prediction.

Stresstesting.   Taking this one step further, one can use statistical simulation for constructing stressmarks, or synthetic benchmarks that stress the processor for a particular metric, e.g., max power consumption, max temperature, max peak power, etc. Current practice is to manually construct stressmarks which is both tedious and time-consuming. An automated stressmark building framework can reduce this overhead and cost. This can be done by integrating statistical simulation in a tuning framework that explores the workload space (by changing the statistical profile) while searching for the stressmark of interest. Joshi et al. [101] describe such a framework that uses a genetic algorithm to search the workload space and automatically generate stressmarks for various stress conditions.

Program behavior characterization.   Another interesting application for statistical simulation is program characterization. When validating the statistical simulation methodology in general and the characteristics included in the statistical profile more in particular, it becomes clear which program characteristics must be included in the profile for attaining good accuracy. That is, this validation process distinguishes program characteristics that influence performance from those that do not.

Workload characterization.   Given that the statistical profile captures the most significant program characteristics, it can be viewed of as an abstract workload model or a concise signature of the workload's execution behavior [46]. In other words, one can compare workloads by comparing their respective statistical profiles.

Large system evaluation.   Finally, current state-of-the-art in statistical simulation addresses the time-consuming simulation problem of single-core and multi-core processors. However, for larger systems containing several processors, such as multi-chip servers, clusters of computers, datacenters, etc., simulation time is even a bigger challenge. Statistical simulation may be an important and interesting approach for such large systems.

7.3   SINGLE-THREADED WORKLOADS

We now describe the current state-of-the-art in statistical simulation, and we do that in three steps with each step considering a progressively more difficult workload type, going from single-threaded, to multi-program and multi-threaded workloads. This section considers single-threaded workloads; the following sections discuss statistical simulation for multi-program and multi-threaded workloads.

Figure 7.2 illustrates statistical simulation for single-threaded workloads in more detail.

7.3.1   STATISTICAL PROFILING

Statistical profiling takes a workload (program trace or binary) as input and computes a set of characteristics. This profiling step is done using a specialized simulator. This could be a specialized trace-driven simulator, a modified execution-driven simulator, or the program binary could even be instrumented to collect the statistics during a program run (see Chapter 5 for a discussion on different forms of simulation). The statistical profile captures the program execution for a specific program input, i.e., the trace that serves as input was collected for a specific input, or the (instrumented) binary is given a specific input during statistics collection. Statistical profiling makes a distinction between characteristics that are microarchitecture-independent versus microarchitecture-dependent.

Microarchitecture-independent characteristics.   A minimal statistical model would collect the instruction mix, i.e., percentage loads, stores, branches, integer operations, floating-point operations,

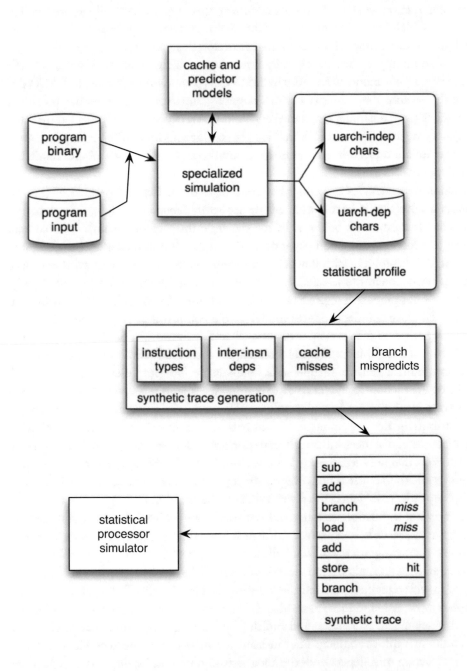

Figure 7.2: Statistical simulation framework for single-threaded workloads.

etc. in the dynamic instruction stream. The number of instruction types is typically limited — the goal is only to know to which functional unit to steer the instruction during simulation. In addition, it includes a distribution that characterizes the inter-instruction dependences (through both registers and memory). Current approaches are typically limited to modeling read-after-write (RAW) dependencies, i.e., they do not model write-after-write (WAW) and write-after-read (WAR) dependencies, primarily because these approaches are targeting superscalar out-of-order processor architectures in which register renaming removes WAW and WAR dependences (provided that sufficient rename registers are available) and in which load bypassing and forwarding is implemented. Targeting statistical simulation towards simpler in-order architectures is likely to require modeling WAR and WAW dependences as well.

Inter-instruction dependencies can either be modeled as downstream dependences (i.e., an instruction produces a value that is later consumed) or upstream dependences (i.e., an instruction consumes a value that was produced before it in the dynamic instruction stream). Downstream dependences model whether an instruction depends on the instruction immediately before it, two instructions before it, etc. A problem with downstream dependencies occurs during synthetic trace generation when an instruction at the selected dependence distance does not produce a value, e.g., a store or a branch does not produce a register value. One solution is to go back to the distribution and try again [47; 152]. Upstream dependences model have a complementary problem. Upstream dependences model whether an instruction at distance d in the future will be dependent on the currently generated instruction. This could lead to situations where instructions have more (or less) incoming dependences than they have input operands. One solution is to simply let this happen [149].

In theory, inter-instruction dependence distributions have infinite size (or at least as large as the instruction trace size) because long dependence distances may exist between instructions. Fortunately, the distributions that are stored on disk as part of a statistical profile can be limited in size because long dependence distances will not affect performance. For example, a RAW dependence between two instructions that are further away from each other in the dynamic instruction stream than the processor's reorder buffer size is not going to affect performance anyway, so there is no need to model such long dependences. Hence, the distribution can be truncated.

The initial statistical simulation methods did not model any notion of control flow behavior [23; 47]: the program characteristics in the statistical profile are simply aggregate metrics, averaged across all instructions in the program execution. Follow-on work started modeling control flow behavior in order to increase the accuracy of statistical simulation. Oskin et al. [152] propose the notion of a control flow graph with transition probabilities between the basic blocks. However, the program characteristics were not correlated to these basic blocks, i.e., they use aggregate statistics, and hence did not increase accuracy much. Nussbaum and Smith [149] correlate program characteristics to basic block size in order to improve accuracy. They measure a distribution of the dynamic basic block size and compute instruction mix, dependence distance distributions and locality events (which we describe next) conditionally dependent on basic block size.

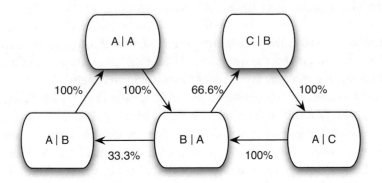

Figure 7.3: An example statistical flow graph (SFG).

Eeckhout et al. [45] propose the statistical flow graph (SFG) which models the control flow using a Markov chain; various program characteristics are then correlated to the SFG. The SFG is illustrated in Figure 7.3 for the **AABAABCABC** example basic block sequence. In fact, this example shows a first-order SFG because it shows transition probabilities between nodes that represent a basic block along with the basic block executed immediately before it. (Extending towards higher-order SFGs is trivial.) The nodes here are A|A, B|A, A|B, etc.: A|B refers to the execution of basic block A given that basic block B was executed immediately before basic block A. The percentages at the edges represent the transition probabilities between the nodes. For example, there is a 33.3% and 66.6% probability to execute basic block A and C, respectively, after having executed basic block A and then B (see the outgoing edges from B|A node). The idea behind the SFG and the reason why it improves accuracy is that, by correlating program characteristics along the SFG, it models execution path dependent program behavior.

All of the characteristics discussed so far are independent of any microarchitecture-specific organization. In other words, these characteristics do not rely on assumptions related to processor issue width, window size, number of ALUs, instruction execution latencies, etc. They are, therefore, called microarchitecture-independent characteristics.

Microarchitecture-dependent characteristics.   In addition to the above characteristics, we also measure a number of characteristics that are related to locality events, such as cache and branch predictor miss rates. Locality events are hard to model in a microarchitecture-independent way. Therefore, a pragmatic approach is taken and characteristics for specific branch predictors and specific cache configurations are computed through specialized cache and branch predictor simulation. Note that although this approach requires the simulation of the complete program execution for specific branch predictors and specific cache structures, this does not limit its applicability. In particular, multiple cache configuration can be simulated in parallel using a single-pass algorithm [83; 135]. An alternative approach is to leverage cache models that predict miss rates. For example, Berg and Hagersten [14] propose a simple statistical model that estimates cache miss rates for a range of caches based on a distribution of reuse latencies (the number of memory references between two references to the same memory location). In other words, rather than computing the miss rates through specialized simulation, one could also use models to predict the miss rates. While this may reduce the time needed to collect the statistical profile, it may come at the cost of increased inaccuracy.

The locality events captured in the initial frameworks are fairly simple [23; 47; 149; 152]. For the branches, it comprises (i) the probability for a taken branch, which puts a limit on the number of taken branches that are fetched per clock cycle; (ii) the probability for a fetch redirection, which corresponds to a branch target misprediction in conjunction with a correct taken/not-taken prediction for conditional branches — this typically results in (a) bubble(s) in the pipeline; and (iii) the probability for a branch misprediction: a BTB miss for indirect branches and a taken/not-taken misprediction for conditional branches. The cache statistics typically consist of the following probabilities: (i) the L1 instruction cache miss rate, (ii) the L2 cache miss rate due to instructions, (iii) the L1 data cache miss rate, (iv) the L2 cache miss rate due to data accesses only, (v) the instruction translation lookaside buffer (I-TLB) miss rate, and (vi) the data translation lookaside buffer (D-TLB) miss rate. (Extending to additional levels of caches and TLBs is trivial.)

At least two major problems with these simple models occur during synthetic trace simulation. First, it does not model delayed hits or hits to outstanding cache lines — there are only hits and misses. The reason is that statistical profiling only sees hits and misses because it basically is a (specialized) functional simulation that processes one memory reference at a time, and it does not account for timing effects. As a result, a delayed hit is modeled as a cache hit although it should see the latency of the outstanding cache line. Second, it does not model the number of instructions in the dynamic instruction stream between misses. However, this has an important impact on the available memory-level parallelism (MLP). Independent long-latency load misses that are close enough to each other in the dynamic instruction stream to make it into the reorder buffer together, potentially overlap their execution, thereby exposing MLP. Given the significant impact of MLP on out-of-order processor performance [31; 102], a performance model lacking adequate MLP modeling may yield large performance prediction errors. Genbrugge and Eeckhout [71] address these problems by modeling the cache statistics conditionally dependent on the (global) cache hit/miss history — by doing so, they model the correlation between cache hits and misses which enables modeling MLP effects. They also collect cache line reuse distributions (i.e., number of memory references between two references to the same cache lines) in order to model delayed hits to outstanding cache lines.

7.3.2   SYNTHETIC TRACE GENERATION

The second step in the statistical simulation methodology is to generate a synthetic trace from the statistical profile. The synthetic trace generator takes as input the statistical profile and produces a synthetic trace that is fed into the statistical simulator. Synthetic trace generation uses random number generation for generating a number in [0,1]; this random number is then used with the inverse cumulative distribution function to determine a program characteristic, see Figure 7.4. The

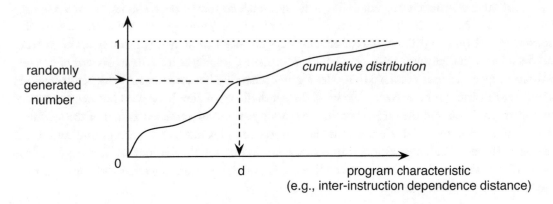

Figure 7.4: Synthetic trace generation: determining a program characteristic through random number generation.

synthetic trace is a linear sequence of synthetically generated instructions. Each instruction has an instruction type, a number of source operands, an inter-instruction dependence for each register input (which denotes the producer for the given register dependence in case of downstream dependence distance modeling), I-cache miss info, D-cache miss info (in case of a load), and branch miss info (in case of a branch). The locality miss events are just labels in the synthetic trace describing whether the load is an L1 D-cache hit, L2 hit or L2 miss and whether the load generates a TLB miss. Similar labels are assigned for the I-cache and branch miss events.

7.3.3   SYNTHETIC TRACE SIMULATION

Synthetic trace simulation is very similar to trace-driven simulation of a real program trace. The simulator takes as input a (synthetic) trace and simulates the timing for each instruction as it moves through the processor pipeline: a number of instructions are fetched, decoded and dispatched each cycle; instructions are issued to a particular functional unit based on functional unit availability and depending on instruction type and whether its dependences have been resolved. When a mispredicted branch is fetched, the pipeline is filled with instructions (from the synthetic trace) as if they were from the incorrect path. When the branch gets executed, the synthetic instructions down the pipeline are squashed and synthetic instructions are re-fetched (and now they are considered from the correct path). In case of an I-cache miss, the fetch engine stops fetching instructions for a number of cycles. The number of cycles is determined by whether the instruction causes an L1 I-cache miss, an L2 cache miss or a D-TLB miss. For a load, the latency will be determined by whether this load is an L1 D-cache hit, an L1 D-cache miss, an L2 cache miss, or a D-TLB miss. For example, in case of an L2 miss, the access latency to main memory is assigned. For delayed hits, the latency assigned is the remaining latency for the outstanding cache line.

Although synthetic trace simulation is fairly similar to trace-driven simulation of a real program trace, the simulator itself is less complex because there are many aspects a trace-driven simulator needs to model that a synthetic trace simulator does not need to model. For example, the instruction decoder in the synthetic trace simulator is much less complex as it needs to discern only a few instruction types. It does not need to model register renaming because the RAW dependencies are already part of the synthetic trace. Likewise, the synthetic trace simulator does not need to model the branch predictor and the caches because the locality events are modeled as part of the synthetic trace. The fact that the synthetic trace simulator is less complex has two important implications. It is easier to develop, and, in addition, it runs faster, i.e., evaluation time is shorter. This is part of the trade-off that statistical simulation offers: it reduces evaluation time as well as development time while incurring some inaccuracy.

7.4   MULTI-PROGRAM WORKLOADS

The approach to statistical simulation as described above models cache behavior through miss statistics. Although this is appropriate for single-core processors, it is inadequate for modeling multicore processors with shared resources, such as shared L2 and/or L3 caches, shared off-chip bandwidth, interconnection network and main memory. Co-executing programs on a chip multiprocessor affect each other's performance through conflicts in these shared resources, and the level of interaction between co-executing programs is greatly affected by the microarchitecture — the amount of interaction can be (very) different across microarchitectures. Hence, cache miss rates profiled from single-threaded execution are unable to model conflict behavior in shared resources when co-executing multiple programs on multicore processors. Instead, what we need is a model that is independent of the memory hierarchy so that conflict behavior among co-executing programs can be derived during multicore simulation of the synthetic traces.

Genbrugge and Eeckhout [72] propose two additional program characteristics, namely, the *cache set profile* and the *per-set LRU stack depth profile* for the largest cache of interest (i.e., the largest cache the architect is interested in during design space exploration). The cache set profile basically is a distribution that captures the fraction of accesses to each set in the cache. The per-set LRU stack depth profile stores the fraction of accesses to each position in the LRU stack. These program characteristics solve two problems: it makes the cache locality profile largely microarchitecture-independent — it only depends on cache line size — which enables estimating cache miss rates for smaller sized caches (i.e., caches with smaller associativity and/or fewer sets). (The underlying mechanism is very similar to the MSI checkpointing mechanism explained in Section 6.3, see also Figure 6.5.) In addition, the cache set and per-set LRU stack depth profiles allows for modeling conflict behavior in shared caches. The only limitation though is that synthetic trace simulation requires that the shared cache(s) is (are) simulated in order to figure out conflict misses among co-executing programs/threads.

A critical issue to multicore processor performance modeling is that the synthetic trace should accurately capture the program's time-varying execution behavior. The reason is that a program's behavior is greatly affected by the behavior of its co-executing program(s), i.e., the relative progress of a program is affected by the conflict behavior in the shared resources [24; 188]. In particular, cache misses induced through cache sharing may slow down a program's execution, which in its turn may result in different sharing behavior. A simple but effective solution is to model a program's time-varying behavior by dividing the program trace in a number of instruction intervals and generating a synthetic mini-trace for each of these instruction intervals [72]. Coalescing the mini-traces then yields a synthetic trace that captures the original program's time-varying execution behavior.

7.5   MULTI-THREADED WORKLOADS

Nussbaum and Smith [150] extended the statistical simulation methodology towards symmetric multiprocessor (SMP) systems running multi-threaded workloads. This requires modeling inter-thread synchronization and communication. More specifically, they model cache coherence events, sequential consistency events, lock accesses and barrier distributions. For modeling cache coherence events and sequential consistency effects, they model whether a store writes to a cache line that it does not own, in which case it will not complete until the bus invalidation has reached the address bus. Also, they model whether a sequence of consecutive stores access private versus shared memory pages. Consecutive stores to private pages can be sent to memory when their input registers are available; consecutive stores to shared pages can only be sent to memory if the invalidation of the previous store has reached the address bus in order to preserve sequential consistency.

Lock accesses are modeled through acquire and release instructions in the statistical profile and synthetic trace. More specifically, for an architecture that implements critical sections through load-linked and store-conditional instructions, the load-linked instruction is retried until it finds the lock variable is clear. It then acquires the lock through a store-conditional to the same lock variable. If the lock variable has been invalidated since the load-linked instruction, this indicates that another thread entered the critical section first. Statistical simulation models all instructions executed between the first load-linked — multiple load-linked instructions may need to be executed before it sees the lock variable is clear — and the successful store-conditional as a single acquire instruction. When a thread exits the critical section, it releases the lock by storing a zero in the lock variable through a conventional store instruction; this is modeled through a single release instruction in statistical simulation. A distribution of lock variables is also maintained in order to be able to discern different critical sections and have different probabilities for entering each critical section. During statistical simulation, a random number along with the lock variable distribution then determines which critical section a thread is entering. Separate statistical profiles are computed for code executed outside versus inside critical sections.

Finally, modeling barrier synchronization is done by counting the number of instructions executed per thread between two consecutive barriers. During synthetic trace generation, the number of instructions between barriers is then scaled down proportionally to the number of instructions executed during detailed execution relative to the number of instructions during statistical simulation.

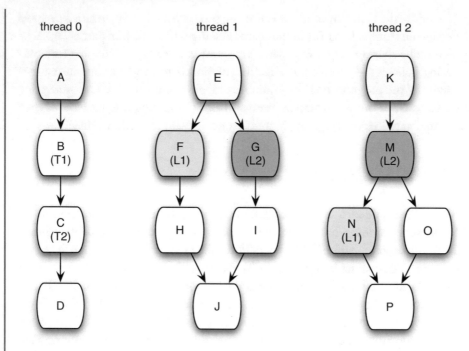

Figure 7.5: Synchronized statistical flow graph (SSFG) illustrative example.

The idea is to scale down the amount of work done between barriers in proportion to the length of the statistical simulation.

A limitation of the Nussaum and Smith approach is that many of the characteristics are microarchitecture-dependent; hence, the method requires detailed simulation instead of specialized functional simulation during statistical profiling. Therefore, Hughes and Li [87] propose the concept of a synchronized statistical flow graph (SSFG), which is a function only of the program under study and not the microarchitecture. The SSFG is illustrated in Figure 7.5. Thread $T0$ is the main thread, and $T1$ and $T2$ are two child threads. There is a separate SFG for each thread, and thread spawning is marked explicitly in the SSFG, e.g., $T1$ is spawned in node B of $T0$ and $T2$ is spawned in node C of $T0$. In addition, the SSFG also models critical sections and for each critical section which lock variable it accesses. For example, node F in $T1$ accesses the same critical section as node N in $T2$; they both access the same lock variable $L1$.

7.6   OTHER WORK IN STATISTICAL MODELING

There exist a number of other approaches that use statistics in performance modeling. Noonburg and Shen [148] model a program execution as a Markov chain in which the states are determined by the microarchitecture and the transition probabilities by the program. This approach works well for simple in-order architectures because the state space is relatively small. However, extending this approach towards superscalar out-of-order architectures explodes the state space and results in a far too complex Markov chain.

Iyengar et al. [91] present SMART[1], which generates representative synthetic traces based on the concept of a fully qualified basic block. A fully qualified basic block is a basic block along with its context of preceding basic blocks. Synthetic traces are generated by coalescing fully qualified basic blocks of the original program trace so that they are representative for the real program trace while being much shorter. The follow-on work in [90] shifted the focus from the basic block granularity to the granularity of individual instructions. A fully qualified instruction is determined by its preceding instructions and their behavior, i.e., instruction type, I-cache hit/miss, and, if applicable, D-cache and branch misprediction behavior. As a result, SMART makes a distinction between two fully qualified instructions having the same sequence preceding instructions, except that, the behavior may be slightly different, e.g., in one case, a preceding instruction missed in the cache, whereas in the other case it did not. As a result, collecting all these fully qualified instructions during statistical profiling results in a huge amount of data to be stored in memory. For some benchmarks, the authors report that the amount of memory that is needed can exceed the available memory in a machine, so that some information needs to be discarded from the graph.

Recent work also focused on generating synthetic benchmarks rather than synthetic traces. Hsieh and Pedram [86] generate a fully functional program from a statistical profile. However, all the characteristics in the statistical profile are microarchitecture-dependent, which makes this technique useless for microprocessor design space explorations. Bell and John [13] generate short synthetic benchmarks using a collection of microarchitecture-independent and microarchitecture-dependent characteristics similar to what is done in statistical simulation as described in this chapter. Their goal is performance model validation of high-level architectural simulators against RTL-level cycle-accurate simulators using small but representative synthetic benchmarks. Joshi et al. [100] generate synthetic benchmarks based on microarchitecture-independent characteristics only, and they leverage that framework to automatically generate stressmarks (i.e., synthetic programs that maximize power consumption, temperature, etc.), see [101].

A couple other research efforts model cache performance in a statistical way. Berg and Hagersten [14] propose light-weight profiling to collect a memory reference reuse distribution at low overhead and then estimate cache miss rates for random-replacement caches. Chandra et al. [24] propose performance models to predict the impact of cache sharing on co-scheduled programs. The output provided by the performance model is an estimate of the number of extra cache misses for each thread due to cache sharing.

---

[1] SMART should not be confused with SMARTS [193; 194], which is a statistical sampling approach, as described in Chapter 6.

CHAPTER 8

Parallel Simulation and Hardware Acceleration

Computer architects typically have to run a large number of simulations when sweeping across the design space to understand performance sensitivity to specific design parameters. They therefore distribute their simulations on a large computing cluster — all of these simulations are independent of each other and are typically run in parallel. Distributed simulation achieves high simulation throughput — the increase in throughput is linear in the number of simulation machines. Unfortunately, this does not reduce the latency of obtaining a single simulation result. This is a severe problem for simulation runs that take days or even weeks to run to completion. Waiting for these simulations to finish is not productive and slows down the entire design process. In practice, individual simulations that finish in a matter of hours (e.g., overnight) or less is preferred. In other words, obtaining a single simulation result quickly is important because it may be a critical result that moves research and development forward.

One way to speed up individual simulation runs is to exploit parallelism. This chapter describes three approaches for doing so: (i) sampled simulation, which distributes sampling units across a cluster of machines, (ii) parallel simulation, which exploits coarse-grain parallelism to map a software simulator on parallel hardware, e.g., a multicore processor, SMP, cluster of machines, etc., and (iii) FPGA-accelerated simulation, which exploits fine-grain parallelism by mapping a simulator on FPGA hardware.

8.1 PARALLEL SAMPLED SIMULATION

As mentioned in Chapter 6, sampled simulation leans well towards parallel simulation, provided that checkpointing is employed for constructing the architecture and microarchitecture starting images for each sampling unit. By loading the checkpoints in different instances of the simulators across a cluster of machines, one can simulate multiple sampling units in parallel, which greatly reduces turnaround time of individual simulation runs. Several researchers explored this avenue, see for example Lauterbach [118] and Nguyen et al. [146]. More recently, Wenisch et al. [190] report a thousand fold reduction in simulation turnaround time through parallel simulation: they distribute thousands of sampling units (along with their checkpoints) across a cluster of machines.

Girbal et al. [75] propose a different approach to parallel simulation: they partition the dynamic instruction stream in so called *chunks*, and these chunks are distributed across multiple machines, see also Figure 8.1 for an illustrative example. In particular, there are as many chunks as there

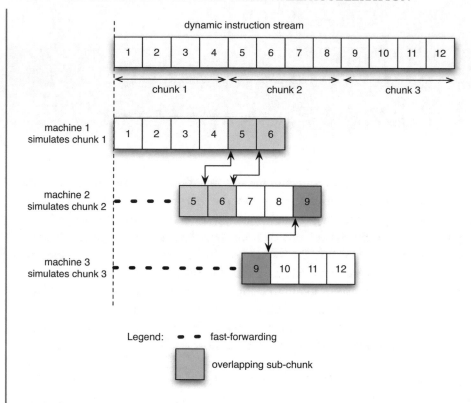

Figure 8.1: Chunk-based parallel simulation.

are machines, and each chunk consists of $1/N$th of the total dynamic instruction stream with N the number of machines. This is different from sampled simulation because sampled simulation simulates sampling units only, in contrast to the Girbal et al. approach which eventually simulates the entire benchmark. Each machine executes the benchmark from the start. The first machine starts detailed simulation immediately; the other machines employ fast-forwarding (or, alternatively, employ checkpointing), and when the beginning of the chunk is reached, they run detailed simulation. The idea now is to continue the simulation of each chunk past its end point (and thus simulate instructions for the next chunk). In other words, there is overlap in simulation load between adjacent machines. By comparing the post-chunk performance numbers against the performance numbers for the next chunk (simulated on the next machine), one can verify whether microarchitecture state has been warmed up on the next machine. (Because detailed simulation of a chunk starts from a cold state on each machine, the performance metrics will be different than for the post-chunk on the previous machine — this is the cold-start problem described in Chapter 6.) Good similarity between the performance numbers will force the simulation to stop on the former machine. The performance numbers at the beginning of each chunk are then discarded and replaced by post-chunk performance numbers from the previous chunk. The motivation for doing so is to compute an overall performance score from performance numbers that were collected from a warmed up state.

8.2 PARALLEL SIMULATION

Most software simulators are single-threaded, which used to be fine because simulation speed benefited from advances in single-thread performance. Thanks to Moore's law and advances in chip technology and microarchitecture, architects were able to improve single-thread performance exponentially. However, this trend has changed over the past decade because of power issues, which has led to the proliferation of chip-multiprocessors (CMPs) or multicore processors. As a result of this multicore trend, it is to be expected that single-thread performance may no longer improve much — and may even deteriorate as one moves towards more power and energy efficient designs. A chip-multiprocessors integrates multiple processor cores on a single chip, and thus a software simulator needs to simulate each core along with their interactions. A simulator that is single-threaded (which is typically the case today) thus is not a scalable solution: whereas Moore's law predicts that the number of cores will double every 18 to 24 months, single-threaded simulator performance will become exponentially slower relative to native processor performance. In other words, the already very wide gap in simulation speed versus native execution speed will grow exponentially as we move forward.

One obvious solution is to parallelize the simulator so that it can exploit the multiple thread contexts in the host multicore processor to speed up simulation. This could potentially lead to a scalable solution in which future generation multicore processors can be simulated on today's multicore processors — simulation would thus scale with each generation of multicore processors. The basic idea is to partition the simulator across multiple simulator threads and have each simulator thread do some part of the simulation work. One possible partitioning is to map each target core onto a separate simulator thread and have one or more simulator threads simulate the shared resources. The parallel simulator itself can be implemented as a single address space shared-memory program, or, alternatively, it can be implemented as a distributed program that uses message passing for communication between the simulator threads.

Parallel simulation is not a novel idea — it has been studied for many years [142; 163] — however, interest has renewed recently given the multicore era [4; 26; 140; 155]. One could even argue that parallel simulation is a necessity if we want simulation to be scalable with advances in multicore processor technology. Interestingly, multicore processors also provide an opportunity for parallel simulation because inter-core communication latencies are much shorter, and there are higher bandwidths between the cores than it used to be case when simulating on symmetric multiprocessors or clusters of machines.

One of the key issues in parallel simulation is to balance accuracy versus speed. Cycle-by-cycle simulation advances one cycle at a time, and thus the simulator threads simulating the target threads need to synchronize every cycle, see Figure 8.2(a). Whereas this is a very accurate approach, its

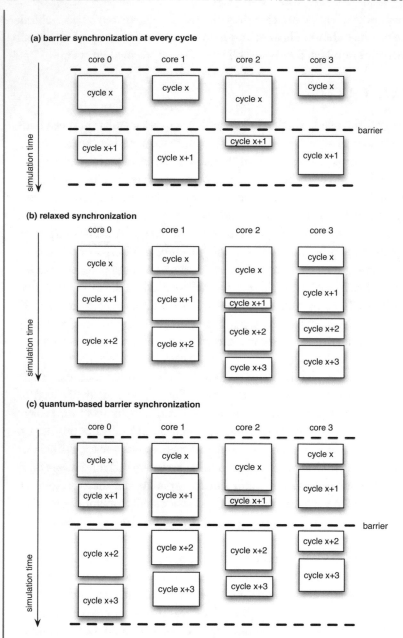

Figure 8.2: Three approaches for synchronizing a parallel simulator that simulates a parallel machine: (a) barrier synchronization at every cycle, (b) relaxed or no synchronization, and (c) quantum-based synchronization.

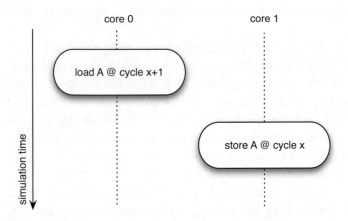

Figure 8.3: Violation of temporal causality due to lack of synchronization: the load sees the old value at memory location A.

performance may not be that great because it requires barrier synchronization at every simulated cycle. If the number of simulator instructions per simulated cycle is low, parallel cycle-by-cycle simulation is not going to yield substantial simulation speed benefits and scalability will be poor.

In order to achieve better simulation performance and scalability, one can relax the cycle-by-cycle condition. In other words, the simulated cores do not synchronize every simulated cycle, which greatly improves simulation speed, see Figure 8.2(b). The downside is that relaxing the synchronization may introduce simulation error. The fundamental reason is that relaxing the cycle-by-cycle condition may lead to situations in which a future event may affect state in the past or a past event does not affect the future — a violation of temporal causality. Figure 8.3 illustrates this. Assume a load instruction to a shared memory address location A is executed at simulated time $x + 1$ and a store instruction to that same address is executed at simulated time x; obviously, the load should see the value written by the store. However, if the load instruction at cycle x would happen to be simulated before the store at cycle $x - 1$, the load would see the old value at memory location A.

There exist two approaches to relax the synchronization imposed by cycle-by-cycle simulation [69]. The optimistic approach takes periodical checkpoints and detects timing violations. When a timing violation is detected, the simulation is rolled back and resumes in a cycle-by-cycle manner until after the timing violation and then switches back to relaxed synchronization. The conservative approach avoids timing violations by processing events only when no other event could possibly affect it. A popular and effective conservative approach is based on barrier synchronization, while relaxing the cycle-by-cycle simulation. The entire simulation is divided into quanta, and each quantum comprises multiple simulated cycles. Quanta are separated through barrier synchronization, see Figure 8.2(c). In other words, simulation threads can advance independently from each other between barriers, and the simulated events become visible to all threads at each barrier. Provided that the time intervals are smaller than the latency for an inter-thread dependence (e.g., to propagate an event from one core to another), temporal causality will be preserved. Hence, quantum-based synchronization achieves cycle-by-cycle simulation accuracy while greatly improving simulation speed compared to cycle-by-cycle simulation. The Wisconsin Wind Tunnel projects [142; 163] implement this approach; the quantum is 100 cycles when simulating shared memory multiprocessors. Chandrasekaran and Hill [25] aim at overcoming the quantum overhead through speculation; Falsafi and Wood [67] leverage multiprogramming to hide the quantum overhead. Falcon et al. [66] propose an adaptive quantum-based synchronization scheme for simulating clusters of machines and the quantum can be as large as 1,000 cycles. When simulating multicore processors, the quantum needs to be smaller because of the relatively small communication latencies between cores: for example, Chidester and George [28] employ a quantum of 12 cycles. A small quantum obviously limits simulation speed. Therefore, researchers are looking into relaxing even further, thereby potentially introducing simulation inaccuracy. Chen et al. [26] study both unbounded slack and bounded slack schemes; Miller et al. [140] study similar approaches. Unbounded slack implies that the slack, or the cycle count difference between two target cores in the simulation, can be as large as the entire simulated execution time. Bounded slack limits the slack to a preset number of cycles, without incurring barrier synchronization.

8.3   FPGA-ACCELERATED SIMULATION

Although parallelizing a simulator helps simulator performance, it is to be expected that the gap between simulation speed and target speed is going to widen, for a number of reasons. First, simulating an n-core processor involves n times more work than simulating a single core for the same simulated time period. In addition, the uncore, e.g., the on-chip interconnection network, tends to grow in complexity as the number of cores increases. As a result, parallel simulation does not yield a linear speedup in the number of host cores, which makes the simulation versus target speed gap grow wider. To make things even worse, it is expected that cache sizes are going to increase as more cores are put on the chip — larger caches reduce pressure on off-chip bandwidth. Hence, ever longer-running benchmarks need to be considered to fully exercise the target machine's caches.

Therefore, interest has grown over the past few years for FPGA-accelerated simulation. The basic idea is to implement (parts of) the simulator on an FPGA (Field Programmable Gate Array). An FPGA is an integrated circuit that can be configured by its user after manufacturing. The main advantage of an FPGA is its rapid turnaround time for new hardware because building a new FPGA design is done quickly: it only requires synthesizing the HDL code (written in a Hardware Description Language, such as VHDL or Verilog) and loading the FPGA code onto the FPGA — much faster than building an ASIC. Also, FPGAs are relatively cheap compared to ASIC designs as well as large SMPs and clusters of servers. In addition, as FPGA density grows with Moore's law, it will be possible to implement more and more processor cores in a single FPGA and hence keep on benefiting from advances in technology, i.e., it will be possible to design next-generation systems on today's technology. Finally, current FPGA-accelerated simulators are substantially faster than software simulators, and they are even fast enough (on the order of tens to hundreds of MIPS, depending on the level of accuracy) to run standard operating systems and complex large scale applications.

FPGA-accelerated simulation thus clearly addresses the evaluation time axis in the simulation diamond (see Chapter 5). Simulation time is greatly reduced compared to software simulation by leveraging the fine-grain parallelism available in the FPGA: work that is done in parallel on the target machine may also be done in parallel on the FPGA, whereas software simulation typically does this in a sequential way. However, simulator development time may be a concern. Because the simulator needs to be synthesizable to the FPGA, it needs to be written in a hardware description language (HDL) like Verilog or VHDL, or at a higher level of abstraction, for example, in Bluespec. This, for one, is likely to increase simulator development time substantially compared to writing a software simulator in a high-level programming language such as C or C++. In addition, software simulators are easily parameterizable and most often do not need recompilation to evaluate a design with a different configuration. An FPGA-accelerated simulator, on the other hand, may need to rerun the entire FPGA synthesis flow (which may take multiple hours, much longer than recompiling a software simulator, if at all needed) when varying parameters during a performance study. Modular simulation infrastructure is crucial for managing FPGA-accelerated simulator development time.

8.3.1   TAXONOMY

Many different FPGA-accelerated simulation approaches have been proposed by various research groups both in industry and academia. In order to understand how these approaches differ from each other, it is important to classify them in terms of how they operate and yield performance numbers. Joel Emer presents a useful taxonomy on FPGA-accelerated simulation [55] and discerns three flavors:

- A *functional emulator* is a circuit that is functionally equivalent to a target design, but does not provide any insight on any specific design metric. A functional emulator is similar to a functional simulator, except that it is implemented in FPGA hardware instead of software. The key advantage of an FPGA-based functional emulator is that it can execute code at hardware speed (several orders of magnitude faster than software simulation), which allows architects and software developers to run commercial software in a reasonable amount of time.

- A *prototype* (or structural emulator) is a functionally equivalent and logically isomorphic representation of the target design. Logically isomorphic means that the prototype implements the same structures as in the target design, and its timing may be scaled with respect to the target system. Hence, a prototype or structural emulator can be used to project performance. For example, a prototype may be a useful vehicle for making high-level design decisions, e.g., study the scalability of software code and/or architecture proposal.

- A *model* is a representation that is functionally equivalent and logically isomorphic with the target design, such that a design metric of interest (of the target design), e.g., performance, power and/or reliability, can be faithfully quantified. The advantage of a model compared to a prototype is that it allows for some abstraction which simplifies model development and enables modularity and easier evaluation of target design alternatives. The issue of representing time faithfully then requires that a distinction is made between a simulated cycle (of the target system) versus an FPGA cycle.

8.3.2   EXAMPLE PROJECTS

As mentioned before, there are several ongoing projects that focus on FPGA-accelerated simulation. The RAMP (Research Accelerator for Multiple Processors) project [5] is a multi-university collaboration that develops open-source FPGA-based simulators and emulators of parallel architectures. The RAMP infrastructure includes a feature that allows for having the inter-component communication channels run as fast as the underlying FPGA hardware will allow, which enables building a range of FPGA simulators from functional emulators, to prototypes and models. The RAMP project has built a number of prototypes, such as RAMP Red and RAMP Blue, and models such as RAMP Gold. The purpose of these designs ranges from evaluating architectures with transactional memory support, message passing machines, distributed shared memory machines, and manycore processor architectures.

A number of other projects are affiliated with RAMP as well, such as Protoflex, FAST and HAsim. Protoflex from Carnegie Mellon University takes a hybrid simulation approach [32] and implements the functional emulator in hardware to fast-forward between sampling units; the detailed cycle-accurate simulation of the sampling units is done in software. Protoflex also augments the functional emulator with cache simulation (e.g., to keep cache state warm between sampling units). Both FAST and HAsim are cycle-level models that partition the functional and timing parts. A principal difference is in the implementation strategy for the functional part. FAST [30], at the University of Texas at Austin, implements a speculative functional-first simulation strategy, see Section 5.6. A functional simulator generates a trace of instructions that is fed into a timing simulator. The functional simulator is placed in software and the timing simulator is placed on the FPGA. The functional simulator speculates on branch outcomes [30; 178] and the memory model [29], and rolls back to an earlier state when mis-speculation is detected. HAsim [154], a collaboration effort between Intel and MIT, is a timing-directed execution-driven simulator, in the terminology of Section 5.6, which requires tight coupling between the functional and timing models: the functional part performs an action in response to a request from the timing part. The functional and timing models are placed on the FPGA. Rare events, such as system calls, are handled in software, alike what is done in Protoflex.

An earlier study using FPGAs to accelerate simulation was done at Princeton University. Penry et al. [155] employ FPGAs as an accelerator for the modular Liberty simulator.

A current direction of research in FPGA-accelerated simulation is to time-division multiplex multiple models. This allows for simulating more components, e.g., cores, on a single FPGA. Protoflex does this for functional emulation; RAMP Gold and HAsim multiplex the timing models.

CHAPTER 9

Concluding Remarks

This book covered a wide spectrum of performance evaluation topics, ranging from performance metrics to workload design, to analytical modeling and various simulation acceleration techniques such as sampled simulation, statistical simulation, and parallel and hardware-accelerated simulation. However, there are a number of topics that this book did not cover. We will now briefly discuss a couple of these.

9.1 TOPICS THAT THIS BOOK DID NOT COVER (YET)

9.1.1 MEASUREMENT BIAS

Mytkovicz et al. [160] study the effect of measurement bias on performance evaluation results. Measurement bias occurs when the experimental setup is biased. This means that when comparing two design alternatives, the experimental setup favors one of the two alternatives so that the benefit of one alternative may be overstated; it may even be the case that the experiment states that one alternative outperforms the other one, even if it is not. In other words, the conclusion may be a result of a biased experimental setup and not because of one alternative being better than the other.

The significance of the work by Mytkovicz et al. is that they have shown that seemingly innocuous aspects of an experimental setup can lead systems researchers to draw incorrect conclusions in practical studies (including simulation studies), and thus, measurement bias cannot be ignored. Mytkowicz et al. show that measurement bias is commonplace across architectures, compilers and benchmarks. They point out two sources of measurement bias. The first source is due to the size of the UNIX environment, i.e., the number of bytes required to store the environment variables. The environment size may have an effect on the layout of the data stored on the stack, which in its turn may affect performance-critical characteristics such as cache and TLB behavior. For example, running the same simulation in different directories may lead to different performance numbers because the directory name affects the environment size. The second source is due to the link order, i.e., the order of object files (with .o extension) given to the linker. The link order may affect code and data layout, and thus overall performance. The fact that there may be other unknown sources of measurement bias complicates the evaluation process even further.

Mytkovicz et al. present two possible solutions. The first solution is to randomize the experimental setup, run each experiment in a different experimental setup, and summarize the performance results across these experimental setups using statistical methods (i.e., compute average and confidence intervals). The obvious downside is that experimental setup randomization drastically increases the number of experiments that need to be run. The second solution is to establish confidence that the outcome is valid even in the presence of measurement bias.

9.1.2   DESIGN SPACE EXPLORATION

We briefly touched upon the fact that computer designers need to explore a large design space when developing the next generation processor. Also, researchers need to run a huge number of simulations in order to understand the performance sensitivity of a new feature in relation to other microarchitecture design parameters.

A common approach is the one-parameter-at-a-time method, which keeps all microarchitecture parameters constant while varying one parameter of interest. The pitfall may be though that some constant parameter may introduce measurement bias, i.e., the effect of the parameter of interest may be overstated or understated. Also, using the one-parameter-at-a-time approach during design space exploration may lead to a suboptimal end result, i.e., the search process may end up in a local optimum, which may be substantially worse than the global optimum.

Yi et al. [160] propose the Plackett and Burman design of experiment to identify the key microarchitecture parameters in a given design space while requiring a small number of simulations only. (We discussed the Plackett and Burman design of experiment in Chapter 3 as a method for finding similarities across benchmarks; however, the method was originally proposed for identifying key microarchitecture parameters.) The end result of a Plackett and Burman design of experiment is a ranking of the most significant microarchitecture parameters. This ranking can be really helpful during design space exploration: it guides the architects to first explore along the most significant microarchitecture parameters before considering the other parameters. This strategy gives higher confidence of finding a good optimum with a limited number of simulations.

Eyerman et al. [62] consider various search algorithms to explore the microarchitecture design space, such as tabu search and genetic algorithms. These algorithms were found to be better at avoiding local optima than the one-parameter-at-a-time strategy at the cost of more simulations. They also propose a two-phase simulation approach in which statistical simulation is first used to identify a region of interest which is then further explored through detailed simulation.

Karkhanis and Smith [104] and Lee and Brooks [120] use analytical models to exhaustively explore a large design space. Because an analytical model is very fast (i.e., a performance prediction is obtained instantaneously), a large number of design points can be explored in an affordable amount of time.

Design space exploration is also crucial in embedded system design. The constraints on the time-to-market are tight, and there is great need for efficient exploration techniques. Virtual prototyping, high-abstraction models, and transaction-level modeling (TLM), which focuses on data transfer functionality rather than implementation, are widely used techniques in embedded system design during early stages of the design cycle.

9.1.3   SIMULATOR VALIDATION

A nagging issue to architectural simulation relates to validation. Performance simulators model a computer architecture at some level of detail, and they typically do not model the target architecture in a cycle-accurate manner; this is especially true for academic simulators. (Although these simulators are often referred to as cycle-accurate simulators, a more appropriate term would probably be cycle-level simulators.) The reason for the lack of validation is threefold. For one, although the high-level design decisions are known, many of the details of contemporary processors are not well described so that academics can re-implement them in their simulators. Second, implementing those details is time-consuming and is likely to severely slowdown the simulation. Third, research ideas (mostly) target future processors for which many of the details are not known anyway, so a generic processor model is a viable option. (This is even true in industry during the early stages of the design cycle.)

Desikan et al. [40] describe the tedious validation process of the `sim-alpha` simulator against the Alpha 21264 processor. They were able to get the mean error between the hardware and the simulator down to less than 2% for a set of microbenchmarks; however, the errors were substantially higher for the SPEC CPU2000 benchmarks (20% on average).

Inspite of the fact that simulator validation is time-consuming and tedious, it is important to validate the simulators against real hardware in order to gain more confidence in the performance numbers that the simulators produce. Again, this is a matter of measurement bias: inadequate modeling of the target system may lead to misleading or incorrect conclusions in practical research studies.

9.2   FUTURE WORK IN PERFORMANCE EVALUATION METHODS

Performance evaluation methods are at the foundation of computer architecture research and development. Hence, it will remain to be an important topic to both academia and industry in the future. Both researchers in academia and industry, as well as practitioners will care about rigorous performance evaluation, because it is one of the key elements that drives research and development forward. There are many challenges ahead of us in performance evaluation.

9.2.1   CHALLENGES RELATED TO SOFTWARE

Software stacks are becoming more and more complex. A modern software stack may consist of a hypervisor (virtual machine monitor or VMM), multiple guest virtual machines each running an operating system, process virtual machines (Java Virtual Machine or Microsoft's .NET framework), middleware, libraries, application software, etc. Whereas simulating application software, like the SPEC CPU benchmarks, is well understood and fairly trivial, simulating more complex workloads, e.g., virtual machines, is much more difficult and is a largely unsolved problem. For example, recent work in Java performance evaluation [17; 18; 48; 74] reveals that Java performance depends on the virtual machine (including the Just-In-Time compiler), the garbage collector, heap size, Java program input, etc. As a result, choosing the right set of system parameters to obtain meaningful simulation results is non-trivial. Moreover, if one is interested in steady-state performance rather than startup performance, a huge number of instructions may need to be simulated.

In addition, given the trend towards multicore and manycore architectures, more and more applications will co-execute on the same processor, and they will share resources, and they will thus affect each other's performance. This implies that we will have to simulate consolidated workloads, and this is going to increase the simulation requirements even further. Moreover, for a given set of applications of interest, the number of possible combinations of these applications (and their starting points) that need to be simulated quickly explodes in the number of cores and applications.

Finally, different applications have different properties. For example, some applications are best-effort applications, whereas others may have (soft) real-time requirements, or need to deliver some level of quality-of-service (QoS) and need to fulfill a service-level agreement (SLA). The mixture of different application types co-executing on the same platform raises the question how to quantify overall system performance and how to compare different systems.

9.2.2 CHALLENGES RELATED TO HARDWARE

The multicore and manycore era poses several profound challenges to performance evaluation methods. Simulating a multicore or manycore processor is fundamentally more difficult than simulating an individual core. As argued before, simulating an n-core processor involves n times more work than simulating a single core for the same simulated time period; in addition, the uncore is likely to grow in complexity. In other words, the gap in simulation speed versus real hardware speed is likely to grow wider. Given Moore's law, it is to be expected that this gap will grow exponentially over time. Scalable solutions that efficiently simulate next-generation multicore processors on today's hardware are a necessity. Parallel simulation is a promising approach; however, balancing simulation speed versus accuracy is a non-trivial issue, and it is unclear whether parallel simulation will scale towards large core counts at realistic levels of accuracy. FPGA-accelerated simulation may be a more scalable solution because it may benefit from Moore's law to implement more and more cores on a single FPGA and thus keep on benefiting from advances in chip technology.

In addition to the challenges related to scale-up (i.e., more and more cores per chip), scale-out (i.e., more and more compute nodes in a datacenter) comes with its own challenges. The trend towards cloud computing in which users access services in 'the cloud', has let to so called warehouse-scale computers or datacenters that are optimized for a specific set of large applications (or Internet services). Setting up a simulation environment to study warehouse-scale computers is far from trivial because of the large scale — a warehouse-scale computers easily hosts thousands of servers — and because the software is complex — multiple layers of software including operating systems, virtual machines, middleware, networking, application software, etc.

Given this trend towards scale-up and scale-out it is inevitable that we will need higher levels of abstraction in our performance evaluation methods. Cycle-accurate simulation will still be needed to fine-tune individual cores; however, in order to be able to simulate large multicore and manycore processors as well as large datacenters and exascale supercomputers, performance models at higher levels of abstraction will be sorely needed. Analytical models and higher-abstraction simulation models (e.g., statistical simulation approaches) will be absolutely necessary in the near future in order to be able to make accurate performance projections.

9.2.3   FINAL COMMENT

This book did provide a — hopefully — comprehensive overview of the current state-of-the-art in computer architecture performance evaluation methods. And, hopefully, this book did provide the required background to tackle the big challenges ahead of us which will require non-trivial advances on multiple fronts, including performance metrics, workloads, modeling, and simulation paradigms and methodologies.

Bibliography

[1] A. Alameldeen and D. Wood. Variability in architectural simulations of multi-threaded workloads. In *Proceedings of the Ninth International Symposium on High-Performance Computer Architecture (HPCA)*, pages 7–18, February 2003. DOI: 10.1109/HPCA.2003.1183520 58, 59, 60

[2] A. R. Alameldeen and D. A. Wood. IPC considered harmful for multiprocessor workloads. *IEEE Micro*, 26(4):8–17, July 2006. DOI: 10.1109/MM.2006.73 6

[3] J. Archibald and J.-L. Baer. Cache coherence protocols: Evaluation using a multiprocessor simulation model. *ACM Transactions on Computer Systems (TOCS)*, 4(4):273–298, November 1986. DOI: 10.1145/6513.6514 81

[4] E. Argollo, A. Falcón, P. Faraboschi, M. Monchiero, and D. Ortega. COTSon: Infrastructure for full system simulation. *SIGOPS Operating System Review*, 43(1):52–61, January 2009. DOI: 10.1145/1496909.1496921 58, 97

[5] Arvind, K. Asanovic, D. Chiou, J. C. Hoe, C. Kozyrakis, S.-L. Lu, M. Oskin, D. Patterson, J. Rabaey, and J. Wawrzynek. RAMP: Research accelerator for multiple processors — a community vision for a shared experimental parallel HW/SW platform. Technical report, University of California, Berkeley, 2005. 102

[6] D. I. August, S. Girbal J. Chang, D. G. Pérez, G. Mouchard, D. A. Penry, O. Temam, and N Vachharajani. UNISIM: An open simulation environment and library for complex architecture design and collaborative development. *IEEE Computer Architecture Letters*, 6(2):45–48, February 2007. DOI: 10.1109/L-CA.2007.12 61

[7] T. Austin, E. Larson, and D. Ernst. SimpleScalar: An infrastructure for computer system modeling. *IEEE Computer*, 35(2):59–67, February 2002. 51, 55

[8] D. A. Bader, Y. Li, T. Li, and V. Sachdeva. BioPerf: A benchmark suite to evaluate high-performance computer architecture on bioinformatics applications. In *Proceedings of the 2005 IEEE International Symposium on Workload Characterization (IISWC)*, pages 163–173, October 2005. DOI: 10.1109/IISWC.2005.1526013 16

[9] K. C. Barr and K. Asanovic. Branch trace compression for snapshot-based simulation. In *Proceedings of the International Symposium on Performance Analysis of Systems and Software (ISPASS)*, pages 25–36, March 2006. 76

[10] K. C. Barr, H. Pan, M. Zhang, and K. Asanovic. Accelerating multiprocessor simulation with a memory timestamp record. In *Proceedings of the 2005 IEEE International Symposium on Performance Analysis of Systems and Software (ISPASS)*, pages 66–77, March 2005. DOI: 10.1109/ISPASS.2005.1430560 76, 78

[11] C. Bechem, J. Combs, N. Utamaphetai, B. Black, R. D. Shawn Blanton, and J. P. Shen. An integrated functional performance simulator. *IEEE Micro*, 19(3):26–35, May/June 1999. DOI: 10.1109/40.768499 55

[12] R. Bedichek. SimNow: Fast platform simulation purely in software. In *Proceedings of the Symposium on High Performance Chips (HOT CHIPS)*, August 2004. 54

[13] R. Bell, Jr. and L. K. John. Improved automatic testcase synthesis for performance model validation. In *Proceedings of the 19th ACM International Conference on Supercomputing (ICS)*, pages 111–120, June 2005. DOI: 10.1145/1088149.1088164 81, 93

[14] E. Berg and E. Hagersten. Fast data-locality profiling of native execution. In *Proceedings of the International Conference on Measurements and Modeling of Computer Systems (SIGMETRICS)*, pages 169–180, June 2005. DOI: 10.1145/1064212.1064232 88, 93

[15] C. Bienia, S. Kumar, J. P. Singh, and K. Li. The PARSEC benchmark suite: Characterization and architectural implications. In *Proceedings of the International Conference on Parallel Architectures and Compilation Techniques (PACT)*, pages 72–81, October 2008. DOI: 10.1145/1454115.1454128 16

[16] N. L. Binkert, R. G. Dreslinski, L. R. Hsu, K. T. Lim, A. G. Saidi, and S. K. Reinhardt. The M5 simulator: Modeling networked systems. *IEEE Micro*, 26(4):52–60, 2006. DOI: 10.1109/MM.2006.82 54, 55, 61

[17] S. M. Blackburn, P. Cheng, and K. S. McKinley. Myths and realities: The performance impact of garbage collection. In *Proceedings of the International Conference on Measurements and Modeling of Computer Systems (SIGMETRICS)*, pages 25–36, June 2004. 105

[18] S. M. Blackburn, R. Garner, C. Hoffmann, A. M. Khan, K. S. McKinley, R. Bentzur, A. Diwan, D. Feinberg, D. Frampton, S. Z. Guyer, M. Hirzel, A. L. Hosking, M. Jump, H. B. Lee, J. Eliot B. Moss, A. Phansalkar, D. Stefanovic, T. VanDrunen, D. von Dincklage, and B. Wiedermann. The dacapo benchmarks: Java benchmarking development and analysis. In *Proceedings of the Annual ACM SIGPLAN Conference on Object-Oriented Programming, Systems, Languages and Applications (OOPSLA)*, pages 169–190, October 2006. 16, 27, 105

[19] P. Bohrer, J. Peterson, M. Elnozahy, R. Rajamony, A. Gheith, R. Rockhold, C. Lefurgy, H. Shafi, T. Nakra, R. Simpson, E. Speight, K. Sudeep, E. Van Hensbergen, and L. Zhang. Mambo: a full system simulator for the PowerPC architecture. *ACM SIGMETRICS Performance Evaluation Review*, 31(4):8–12, March 2004. DOI: 10.1145/1054907.1054910 54

[20] D. C. Burger and T. M. Austin. The SimpleScalar Tool Set. Computer Architecture News, 1997. See also http://www.simplescalar.com for more information. DOI: 10.1145/268806.268810 52, 61

[21] M. Burtscher and I. Ganusov. Automatic synthesis of high-speed processor simulators. In *Proceedings of the 37th IEEE/ACM Symposium on Microarchitecture (MICRO)*, pages 55–66, December 2004. 52, 72

[22] M. Burtscher, I. Ganusov, S. J. Jackson, J. Ke, P. Ratanaworabhan, and N. B. Sam. The VPC trace-compression algorithms. *IEEE Transactions on Computers*, 54(11):1329–1344, November 2005. DOI: 10.1109/TC.2005.186 54

[23] R. Carl and J. E. Smith. Modeling superscalar processors via statistical simulation. In *Workshop on Performance Analysis and its Impact on Design (PAID), held in conjunction with the 25th Annual International Symposium on Computer Architecture (ISCA)*, June 1998. 81, 86, 88

[24] D. Chandra, F. Guo, S. Kim, and Y. Solihin. Predicting inter-thread cache contention on a chip-multiprocessor architecture. In *Proceedings of the Eleventh International Symposium on High Performance Computer Architecture (HPCA)*, pages 340–351, February 2005. 7, 91, 93

[25] S. Chandrasekaran and M. D. Hill. Optimistic simulation of parallel architectures using program executables. In *Proceedings of the Tenth Workshop on Parallel and Distributed Simulation (PADS)*, pages 143–150, May 1996. 100

[26] J. Chen, M. Annavaram, and M. Dubois. SlackSim: A platform for parallel simulation of CMPs on CMPs. *ACM SIGARCH Computer Architecture News*, 37(2):20–29, May 2009. DOI: 10.1145/1577129.1577134 97, 100

[27] X. E. Chen and T. M. Aamodt. Hybrid analytical modeling of pending cache hits, data prefetching, and MSHRs. In *Proceedings of the International Symposium on Microarchitecture (MICRO)*, pages 59–70, December 2008. 46

[28] M. Chidester and A. George. Parallel simulation of chip-multiprocessor architectures. *ACM Transactions on Modeling and Computer Simulation*, 12(3):176–200, July 2002. DOI: 10.1145/643114.643116 100

[29] D. Chiou, H. Angepat, N. A. Patil, and D. Sunwoo. Accurate functional-first multicore simulators. *IEEE Computer Architecture Letters*, 8(2):64–67, July 2009. DOI: 10.1109/L-CA.2009.44 57, 102

[30] D. Chiou, D. Sunwoo, J. Kim, N. A. Patil, W. Reinhart, D. E. Johnson, J. Keefe, and H. Angepat. FPGA-accelerated simulation technologies (FAST): Fast, full-system, cycle-accurate simulators. In *Proceedings of the Annual IEEE/ACM International Symposium on Microarchitecture (MICRO)*, pages 249–261, December 2007. 61, 62, 102

[31] Y. Chou, B. Fahs, and S. Abraham. Microarchitecture optimizations for exploiting memory-level parallelism. In *Proceedings of the 31st Annual International Symposium on Computer Architecture (ISCA)*, pages 76–87, June 2004. 43, 88

[32] E. S. Chung, M. K. Papamichael, E. Nurvitadhi, J. C. Hoe, K. Mai, and B. Falsafi. ProtoFlex: Towards scalable, full-system multiprocessor simulations using FPGAs. *ACM Transactions on Reconfigurable Technology and Systems*, 2(2), June 2009. Article 15. 102

[33] D. Citron. MisSPECulation: Partial and misleading use of SPEC CPU2000 in computer architecture conferences. In *Proceedings of the 30th Annual International Symposium on Computer Architecture (ISCA)*, pages 52–59, June 2003. DOI: 10.1109/ISCA.2003.1206988 17

[34] B. Cmelik and D. Keppel. SHADE: A fast instruction-set simulator for execution profiling. In *Proceedings of the 1994 ACM SIGMETRICS Conference on Measurement and Modeling of Computer Systems*, pages 128–137, May 1994. DOI: 10.1145/183018.183032 51

[35] T. M. Conte, M. A. Hirsch, and W. W. Hwu. Combining trace sampling with single pass methods for efficient cache simulation. *IEEE Transactions on Computers*, 47(6):714–720, June 1998. DOI: 10.1109/12.689650 54, 76

[36] T. M. Conte, M. A. Hirsch, and K. N. Menezes. Reducing state loss for effective trace sampling of superscalar processors. In *Proceedings of the International Conference on Computer Design (ICCD)*, pages 468–477, October 1996. DOI: 10.1109/ICCD.1996.563595 64, 76

[37] H. G. Cragon. *Computer Architecture and Implementation*. Cambridge University Press, 2000. 11

[38] P. Crowley and J.-L. Baer. Trace sampling for desktop applications on Windows NT. In *Proceedings of the First Workshop on Workload Characterization (WWC) held in conjunction with the 31st ACM/IEEE Annual International Symposium on Microarchitecture (MICRO)*, November 1998. 74

[39] P. Crowley and J.-L. Baer. On the use of trace sampling for architectural studies of desktop applications. In *Proceedings of the 1999 ACM SIGMETRICS International Conference on Measurement and Modeling of Computer Systems*, pages 208–209, June 1999. DOI: 10.1145/301453.301573 74

[40] R. Desikan, D. Burger, and S. W. Keckler. Measuring experimental error in microprocessor simulation. In *Proceedings of the 28th Annual International Symposium on Computer Architecture (ISCA)*, pages 266–277, July 2001. DOI: 10.1145/379240.565338 105

[41] C. Dubach, T. M. Jones, and M. F. P. O'Boyle. Microarchitecture design space exploration using an architecture-centric approach. In *Proceedings of the IEEE/ACM Annual International Symposium on Microarchitecture (MICRO)*, pages 262–271, December 2007. 36

[42] P. K. Dubey and R. Nair. Profile-driven sampled trace generation. Technical Report RC 20041, IBM Research Division, T. J. Watson Research Center, April 1995. 70

[43] M. Durbhakula, V. S. Pai, and S. V. Adve. Improving the accuracy vs. speed tradeoff for simulating shared-memory multiprocessors with ILP processors. In *Proceedings of the Fifth International Symposium on High-Performance Computer Architecture (HPCA)*, pages 23–32, January 1999. DOI: 10.1109/HPCA.1999.744317 71

[44] J. Edler and M. D. Hill. Dinero IV trace-driven uniprocessor cache simulator. Available through http://www.cs.wisc.edu/~markhill/DineroIV, 1998. 54

[45] L. Eeckhout, R. H. Bell Jr., B. Stougie, K. De Bosschere, and L. K. John. Control flow modeling in statistical simulation for accurate and efficient processor design studies. In *Proceedings of the 31st Annual International Symposium on Computer Architecture (ISCA)*, pages 350–361, June 2004. 87

[46] L. Eeckhout and K. De Bosschere. Hybrid analytical-statistical modeling for efficiently exploring architecture and workload design spaces. In *Proceedings of the 2001 International Conference on Parallel Architectures and Compilation Techniques (PACT)*, pages 25–34, September 2001. DOI: 10.1109/PACT.2001.953285 84

[47] L. Eeckhout, K. De Bosschere, and H. Neefs. Performance analysis through synthetic trace generation. In *The IEEE International Symposium on Performance Analysis of Systems and Software (ISPASS)*, pages 1–6, April 2000. 86, 88

[48] L. Eeckhout, A. Georges, and K. De Bosschere. How Java programs interact with virtual machines at the microarchitectural level. In *Proceedings of the 18th Annual ACM SIGPLAN Conference on Object-Oriented Programming, Languages, Applications and Systems (OOPSLA)*, pages 169–186, October 2003. 27, 105

[49] L. Eeckhout, Y. Luo, K. De Bosschere, and L. K. John. BLRL: Accurate and efficient warmup for sampled processor simulation. *The Computer Journal*, 48(4):451–459, May 2005. DOI: 10.1093/comjnl/bxh103 75

[50] L. Eeckhout, S. Nussbaum, J. E. Smith, and K. De Bosschere. Statistical simulation: Adding efficiency to the computer designer's toolbox. *IEEE Micro*, 23(5):26–38, Sept/Oct 2003. DOI: 10.1109/MM.2003.1240210 81

[51] L. Eeckhout, J. Sampson, and B. Calder. Exploiting program microarchitecture independent characteristics and phase behavior for reduced benchmark suite simulation. In *Proceedings of the 2005 IEEE International Symposium on Workload Characterization (IISWC)*, pages 2–12, October 2005. DOI: 10.1109/IISWC.2005.1525996 27, 70

[52] L. Eeckhout, H. Vandierendonck, and K. De Bosschere. Designing workloads for computer architecture research. *IEEE Computer*, 36(2):65–71, February 2003. 27

[53] L. Eeckhout, H. Vandierendonck, and K. De Bosschere. Quantifying the impact of input data sets on program behavior and its applications. *Journal of Instruction-Level Parallelism*, 5, February 2003. http://www.jilp.org/vol5. 18, 25

[54] M. Ekman and P. Stenström. Enhancing multiprocessor architecture simulation speed using matched-pair comparison. In *Proceedings of the 2005 IEEE International Symposium on Performance Analysis of Systems and Software (ISPASS)*, pages 89–99, March 2005. DOI: 10.1109/ISPASS.2005.1430562 70, 79

[55] J. Emer. Accelerating architecture research. IEEE International Symposium on Performance Analysis of Systems and Software (ISPASS), April 2009. Keynote address. 1, 101

[56] J. Emer. Eckert-Mauchly Award acceptance speech. June 2009. 1

[57] J. Emer, P. Ahuja, E. Borch, A. Klauser, C.-K. Luk, S. Manne, S. S. Mukherjee, H. Patil, S. Wallace, N. Binkert, R. Espasa, and T. Juan. Asim: A performance model framework. *IEEE Computer*, 35(2):68–76, February 2002. 55, 56, 61

[58] J. Emer, C. Beckmann, and M. Pellauer. AWB: The Asim architect's workbench. In *Proceedings of the Third Annual Workshop on Modeling, Benchmarking and Simulation (MoBS), held in conjunction with ISCA*, June 2007. 61

[59] J. S. Emer and D. W. Clark. A characterization of processor performance in the VAX-11/780. In *Proceedings of the International Symposium on Computer Architecture (ISCA)*, pages 301–310, June 1984. 7

[60] P. G. Emma. Understanding some simple processor-performance limits. *IBM Journal of Research and Development*, 41(3):215–232, May 1997. DOI: 10.1147/rd.413.0215 6

[61] S. Eyerman and L. Eeckhout. System-level performance metrics for multi-program workloads. *IEEE Micro*, 28(3):42–53, May/June 2008. DOI: 10.1109/MM.2008.44 8, 11

[62] S. Eyerman, L. Eeckhout, and K. De Bosschere. Efficient design space exploration of high performance embedded out-of-order processors. In *Proceedings of the 2006 Conference on Design Automation and Test in Europe (DATE)*, pages 351–356, March 2006. 104

[63] S. Eyerman, L. Eeckhout, T. Karkhanis, and J. E. Smith. A performance counter architecture for computing accurate CPI components. In *Proceedings of The Twelfth International Conference on Architectural Support for Programming Languages and Operating Systems (ASPLOS)*, pages 175–184, October 2006. 45

[64] S. Eyerman, L. Eeckhout, T. Karkhanis, and J. E. Smith. A mechanistic performance model for superscalar out-of-order processors. *ACM Transactions on Computer Systems (TOCS)*, 27(2), May 2009. 38, 44, 45

[65] S. Eyerman, James E. Smith, and L. Eeckhout. Characterizing the branch misprediction penalty. In *IEEE International Symposium on Performance Analysis of Systems and Software (ISPASS)*, pages 48–58, March 2006. 40

[66] A. Falcón, P. Faraboschi, and D. Ortega. An adaptive synchronization technique for parallel simulation of networked clusters. In *Proceedings of the IEEE International Symposium on Performance Analysis of Systems and Software (ISPASS)*, pages 22–31, April 2008. DOI: 10.1109/ISPASS.2008.4510735 100

[67] B. Falsafi and D. A. Wood. Modeling cost/performance of a parallel computer simulator. *ACM Transactions on Modeling and Computer Simulation (TOMACS)*, 7(1):104–130, January 1997. DOI: 10.1145/244804.244808 100

[68] P. J. Fleming and J. J. Wallace. How not to lie with statistics: The correct way to summarize benchmark results. *Communications of the ACM*, 29(3):218–221, March 1986. DOI: 10.1145/5666.5673 11

[69] R. M. Fujimoto. Parallel discrete event simulation. *Communications of the ACM*, 33(10):30–53, October 1990. DOI: 10.1145/84537.84545 99

[70] R. M. Fujimoto and W. B. Campbell. Direct execution models of processor behavior and performance. In *Proceedings of the 19th Winter Simulation Conference*, pages 751–758, December 1987. 71

[71] D. Genbrugge and L. Eeckhout. Memory data flow modeling in statistical simulation for the efficient exploration of microprocessor design spaces. *IEEE Transactions on Computers*, 57(10):41–54, January 2007. 88

[72] D. Genbrugge and L. Eeckhout. Chip multiprocessor design space exploration through statistical simulation. *IEEE Transactions on Computers*, 58(12):1668–1681, December 2009. DOI: 10.1109/TC.2009.77 90, 91

[73] D. Genbrugge, S. Eyerman, and L. Eeckhout. Interval simulation: Raising the level of abstraction in architectural simulation. In *Proceedings of the International Symposium on High-Performance Computer Architecture (HPCA)*, pages 307–318, January 2010. 45

[74] A. Georges, D. Buytaert, and L. Eeckhout. Statistically rigorous java performance evaluation. In *Proceedings of the Annual ACM SIGPLAN Conference on Object-Oriented Programming, Languages, Applications and Systems (OOPSLA)*, pages 57–76, October 2007. 105

[75] S. Girbal, G. Mouchard, A. Cohen, and O. Temam. DiST: A simple, reliable and scalable method to significantly reduce processor architecture simulation time. In *Proceedings of the 2003 ACM SIGMETRICS International Conference on Measurement and Modeling of Computer Systems*, pages 1–12, June 2003. DOI: 10.1145/781027.781029 95

[76] A. Glew. MLP yes! ILP no! In *ASPLOS Wild and Crazy Idea Session*, October 1998. 43

[77] G. Hamerly, E. Perelman, J. Lau, and B. Calder. SimPoint 3.0: Faster and more flexible program analysis. *Journal of Instruction-Level Parallelism*, 7, September 2005. 70

[78] A. Hartstein and T. R. Puzak. The optimal pipeline depth for a microprocessor. In *Proceedings of the 29th Annual International Symposium on Computer Architecture (ISCA)*, pages 7–13, May 2002. DOI: 10.1109/ISCA.2002.1003557 46

[79] J. W. Haskins Jr. and K. Skadron. Accelerated warmup for sampled microarchitecture simulation. *ACM Transactions on Architecture and Code Optimization (TACO)*, 2(1):78–108, March 2005. DOI: 10.1145/1061267.1061272 75

[80] J. L. Hennessy and D. A. Patterson. *Computer Architecture: A Quantitative Approach*. Morgan Kaufmann Publishers, third edition, 2003. 11

[81] J. L. Henning. SPEC CPU2000: Measuring CPU performance in the new millennium. *IEEE Computer*, 33(7):28–35, July 2000. 17

[82] M. D. Hill and M. R. Marty. Amdahl's law in the multicore era. *IEEE Computer*, 41(7):33–38, July 2008. 31

[83] M. D. Hill and A. J. Smith. Evaluating associativity in CPU caches. *IEEE Transactions on Computers*, 38(12):1612–1630, December 1989. DOI: 10.1109/12.40842 54, 87

[84] S. Hong and H. Kim. An analytical model for a GPU architecture with memory-level and thread-level parallelism awareness. In *Proceedings of the International Symposium on Computer Architecture (ISCA)*, pages 152–163, June 2008. 46

[85] K. Hoste and L. Eeckhout. Microarchitecture-independent workload characterization. *IEEE Micro*, 27(3):63–72, May 2007. DOI: 10.1109/MM.2007.56 20, 23

[86] C. Hsieh and M. Pedram. Micro-processor power estimation using profile-driven program synthesis. *IEEE Transactions on Computer-Aided Design of Integrated Circuits and Systems*, 17(11):1080–1089, November 1998. DOI: 10.1109/43.736182 93

[87] C. Hughes and T. Li. Accelerating multi-core processor design space evaluation using automatic multi-threaded workload synthesis. In *Proceedings of the IEEE International Symposium on Workload Characterization (IISWC)*, pages 163–172, September 2008. DOI: 10.1109/IISWC.2008.4636101 81, 92

[88] C. J. Hughes, V. S. Pai, P. Ranganathan, and S. V. Adve. Rsim: Simulating shared-memory multiprocessors with ILP processors. *IEEE Computer*, 35(2):40–49, February 2002. 55

[89] E. Ipek, S. A. McKee, B. R. de Supinski, M. Schulz, and R. Caruana. Efficiently exploring architectural design spaces via predictive modeling. In *Proceedings of the Twelfth International Conference on Architectural Support for Programming Languages and Operating Systems (ASPLOS)*, pages 195–206, October 2006. 36

[90] V. S. Iyengar and L. H. Trevillyan. Evaluation and generation of reduced traces for benchmarks. Technical Report RC 20610, IBM Research Division, T. J. Watson Research Center, October 1996. 93

[91] V. S. Iyengar, L. H. Trevillyan, and P. Bose. Representative traces for processor models with infinite cache. In *Proceedings of the Second International Symposium on High-Performance Computer Architecture (HPCA)*, pages 62–73, February 1996. DOI: 10.1109/HPCA.1996.501174 70, 93

[92] R. K. Jain. *The Art of Computer Systems Performance Analysis: Techniques for Experimental Design, Measurement, Simulation, and Modeling*. Wiley, 1991. xi

[93] L. K. John. More on finding a single number to indicate overall performance of a benchmark suite. *ACM SIGARCH Computer Architecture News*, 32(4):1–14, September 2004. 11, 12

[94] L. K. John and L. Eeckhout, editors. *Performance Evaluation and Benchmarking*. CRC Press, Taylor and Francis, 2006. xi

[95] E. E. Johnson, J. Ha, and M. B. Zaidi. Lossless trace compression. *IEEE Transactions on Computers*, 50(2):158–173, February 2001. DOI: 10.1109/12.908991 54

[96] R. A. Johnson and D. W. Wichern. *Applied Multivariate Statistical Analysis*. Prentice Hall, fifth edition, 2002. 18, 23

[97] P. J. Joseph, K. Vaswani, and M. J. Thazhuthaveetil. Construction and use of linear regression models for processor performance analysis. In *Proceedings of the 12th International Symposium on High-Performance Computer Architecture (HPCA)*, pages 99–108, February 2006. 32, 34, 35

[98] P. J. Joseph, K. Vaswani, and M. J. Thazhuthaveetil. A predictive performance model for superscalar processors. In *Proceedings of the 39th Annual IEEE/ACM International Symposium on Microarchitecture (MICRO)*, pages 161–170, December 2006. 36

[99] A. Joshi, A. Phansalkar, L. Eeckhout, and L. K. John. Measuring benchmark similarity using inherent program characteristics. *IEEE Transactions on Computers*, 55(6):769–782, June 2006. DOI: 10.1109/TC.2006.85 23, 62

[100] A. M. Joshi, L. Eeckhout, R. Bell, Jr., and L. K. John. Distilling the essence of proprietary workloads into miniature benchmarks. *ACM Transactions on Architecture and Code Optimization (TACO)*, 5(2), August 2008. 93

[101] A. M. Joshi, L. Eeckhout, L. K. John, and C. Isen. Automated microprocessor stressmark generation. In *Proceedings of the International Symposium on High-Performance Computer Architecture (HPCA)*, pages 229–239, February 2008. 84, 93

[102] T. Karkhanis and J. E. Smith. A day in the life of a data cache miss. In *Proceedings of the 2nd Annual Workshop on Memory Performance Issues (WMPI) held in conjunction with ISCA*, May 2002. 42, 43, 88

[103] T. Karkhanis and J. E. Smith. A first-order superscalar processor model. In *Proceedings of the 31st Annual International Symposium on Computer Architecture (ISCA)*, pages 338–349, June 2004. 43, 45

[104] T. Karkhanis and J. E. Smith. Automated design of application specific superscalar processors: An analytical approach. In *Proceedings of the 34th Annual International Symposium on Computer Architecture (ISCA)*, pages 402–411, June 2007. DOI: 10.1145/1250662.1250712 45, 104

[105] K. Keeton, D. A. Patterson, Y. Q. He, R. C. Raphael, and W. E. Baker. Performance characterization of a quad Pentium Pro SMP using OLTP workloads. In *Proceedings of the International Symposium on Computer Architecture (ISCA)*, pages 15–26, June 1998. 15

[106] R. E. Kessler, M. D. Hill, and D. A. Wood. A comparison of trace-sampling techniques for multi-megabyte caches. *IEEE Transactions on Computers*, 43(6):664–675, June 1994. DOI: 10.1109/12.286300 74, 75

[107] S. Kluyskens and L. Eeckhout. Branch predictor warmup for sampled simulation through branch history matching. *Transactions on High-Performance Embedded Architectures and Compilers (HiPEAC)*, 2(1):42–61, January 2007. 76

[108] P. Kongetira, K. Aingaran, and K. Olukotun. Niagara: A 32-way multithreaded SPARC processor. *IEEE Micro*, 25(2):21–29, March/April 2005. DOI: 10.1109/MM.2005.35 7

[109] V. Krishnan and J. Torrellas. A direct-execution framework for fast and accurate simulation of superscalar processors. In *Proceedings of the 1998 International Conference on Parallel Architectures and Compilation Techniques (PACT)*, pages 286–293, October 1998. DOI: 10.1109/PACT.1998.727263 71

[110] T. Kuhn. *The Structure of Scientific Revolutions*. University Of Chicago Press, 1962. 1

[111] B. Kumar and E. S. Davidson. Performance evaluation of highly concurrent computers by deterministic simulation. *Communications of the ACM*, 21(11):904–913, November 1978. DOI: 10.1145/359642.359646 81

[112] T. Lafage and A. Seznec. Choosing representative slices of program execution for microar-chitecture simulations: A preliminary application to the data stream. In *IEEE 3rd Annual Workshop on Workload Characterization (WWC-2000) held in conjunction with the International Conference on Computer Design (ICCD)*, September 2000. 67

[113] S. Laha, J. H. Patel, and R. K. Iyer. Accurate low-cost methods for performance evaluation of cache memory systems. *IEEE Transactions on Computers*, 37(11):1325–1336, November 1988. DOI: 10.1109/12.8699 64

[114] J. R. Larus and E. Schnarr. EEL: Machine-independent executable editing. In *Proceedings of the ACM SIGPLAN Conference on Programming Language Design and Implementation (PLDI)*, pages 291–300, June 1995. 51

[115] J. Lau, E. Perelman, and B. Calder. Selecting software phase markers with code structure analysis. In *Proceedings of the International Symposium on Code Generation and Optimization (CGO)*, pages 135–146, March 2006. DOI: 10.1109/CGO.2006.32 68

[116] J. Lau, J. Sampson, E. Perelman, G. Hamerly, and B. Calder. The strong correlation be-tween code signatures and performance. In *Proceedings of the International Symposium on Performance Analysis of Systems and Software (ISPASS)*, pages 236–247, March 2005. DOI: 10.1109/ISPASS.2005.1430578 68

[117] J. Lau, S. Schoenmackers, and B. Calder. Structures for phase classification. In *Proceedings of the 2004 International Symposium on Performance Analysis of Systems and Software (ISPASS)*, pages 57–67, March 2004. DOI: 10.1109/ISPASS.2004.1291356 68

[118] G. Lauterbach. Accelerating architectural simulation by parallel execution of trace samples. Technical Report SMLI TR-93-22, Sun Microsystems Laboratories Inc., December 1993. 70, 76, 95

[119] B. Lee and D. Brooks. Accurate and efficient regression modeling for microarchitectural performance and power prediction. In *Proceedings of the Twelfth International Conference on Architectural Support for Programming Languages and Operating Systems (ASPLOS)*, pages 185–194, October 2006. 35

[120] B. Lee and D. Brooks. Efficiency trends and limits from comprehensive microarchitec-tural adaptivity. In *Proceedings of the 13th International Conference on Architectural Support for Programming Languages and Operating Systems (ASPLOS)*, pages 36–47, March 2008. DOI: 10.1145/1346281.1346288 31, 36, 104

[121] B. Lee, D. Brooks, Bronis R. de Supinski, M. Schulz, K. Singh, and S. A. McKee. Methods of inference and learning for performance modeling of parallel applications. In *Proceedings of the 12th ACM SIGPLAN Symposium on Principles and Practice of Parallel Programming (PPOPP)*, pages 249–258, March 207. 36

[122] B. Lee, J. Collins, H. Wang, and D. Brooks. CPR: Composable performance regression for scalable multiprocessor models. In *Proceedings of the 41st Annual IEEE/ACM International Symposium on Microarchitecture (MICRO)*, pages 270–281, November 2008. 36

[123] B. C. Lee and D. M. Brooks. Illustrative design space studies with microarchitectural regression models. In *Proceedings of the International Symposium on High Performance Computer Architecture (HPCA)*, pages 340–351, February 2007. 36

[124] C. Lee, M. Potkonjak, and W. H. Mangione-Smith. MediaBench: A tool for evaluating and synthesizing multimedia and communications systems. In *Proceedings of the 30th Annual IEEE/ACM Symposium on Microarchitecture (MICRO)*, pages 330–335, December 1997. 16

[125] K. M. Lepak, H. W. Cain, and M. H. Lipasti. Redeeming IPC as a performance metric for multithreaded programs. In *Proceedings of the International Conference on Parallel Architectures and Compilation Techniques (PACT)*, pages 232–243, September 2003. 59

[126] D. J. Lilja. *Measuring Computer Performance: A Practitioner's Guide*. Cambridge University Press, 2000. DOI: 10.1017/CBO9780511612398 xi, 65

[127] M. H. Lipasti, C. B. Wilkerson, and J. P. Shen. Value locality and load value prediction. In *Proceedings of the International Conference on Architectural Support for Programming Languages and Operating Systems (ASPLOS)*, pages 138–147, October 1996. 30

[128] C.-K. Luk, R. Cohn, R. Muth, H. Patil, A. Klauser, G. Lowney, S. Wallace, V. J. Reddi, and K. Hazelwood. Pin: Building customized program analysis tools with dynamic instrumentation. In *Proceedings of the ACM SIGPLAN Conference on Programming Languages Design and Implementation (PLDI)*, pages 190–200, June 2005. DOI: 10.1145/1065010.1065034 22, 51

[129] K. Luo, J. Gummaraju, and M. Franklin. Balancing throughput and fairness in SMT processors. In *Proceedings of the IEEE International Symposium on Performance Analysis of Systems and Software (ISPASS)*, pages 164–171, November 2001. 9, 10

[130] Y. Luo and L. K. John. Efficiently evaluating speedup using sampled processor simulation. *Computer Architecture Letters*, 4, September 2004. 70

[131] Y. Luo, L. K. John, and L. Eeckhout. SMA: A self-monitored adaptive warmup scheme for microprocessor simulation. *International Journal on Parallel Programming*, 33(5):561–581, October 2005. DOI: 10.1007/s10766-005-7305-9 75

[132] P. S. Magnusson, M. Christensson, Jesper Eskilson, D. Forsgren, G. Hallberg, J. Högberg nad F. Larsson, A. Moestedt, and B. Werner. Simics: A full system simulation platform. *IEEE Computer*, 35(2):50–58, February 2002. 53

[133] M. K. Martin, D. J. Sorin, B. M. Beckmann, M. R. Marty, M. Xu, A. R. Alameldeen, K. E. Moore, M. D. Hill, and D. A. Wood. Multifacet's general execution-driven multiprocessor simulator (GEMS) toolset. *ACM SIGARCH Computer Architecture News*, 33(4):92–99, November 2005. DOI: 10.1145/1105734.1105747 55, 61

[134] J. R. Mashey. War of the benchmark means: Time for a truce. *ACM SIGARCH Computer Architecture News*, 32(4):1–14, September 2004. DOI: 10.1145/1040136.1040137 11, 13

[135] R. L. Mattson, J. Gecsei, D. R. Slutz, and I. L. Traiger. Evaluation techniques for storage hierarchies. *IBM Systems Journal*, 9(2):78–117, June 1970. DOI: 10.1147/sj.92.0078 54, 87

[136] C. J. Mauer, M. D. Hill, and D. A. Wood. Full-system timing-first simulation. In *Proceedings of the 2002 ACM SIGMETRICS Conference on Measurement and Modeling of Computer Systems*, pages 108–116, June 2002. DOI: 10.1145/511334.511349 55, 58

[137] A. M. G. Maynard, C. M. Donnelly, and B. R. Olszewski. Contrasting characteristics and cache performance of technical and multi-user commercial workloads. In *Proceedings of the International Conference on Architectural Support for Programming Languages and Operating Systems (ASPLOS)*, pages 145–156, October 1994. DOI: 10.1145/195473.195524 15

[138] P. Michaud, A. Seznec, and S. Jourdan. Exploring instruction-fetch bandwidth requirement in wide-issue superscalar processors. In *Proceedings of the 1999 International Conference on Parallel Architectures and Compilation Techniques (PACT)*, pages 2–10, October 1999. DOI: 10.1109/PACT.1999.807388 35, 45

[139] D. Mihocka and S. Schwartsman. Virtualization without direct execution or jitting: Designing a portable virtual machine infrastructure. In *Proceedings of the Workshop on Architectural and Microarchitectural Support for Binary Translation, held in conjunction with ISCA*, June 2008. 54

[140] J. E. Miller, H. Kasture, G. Kurian, C. Gruenwald III, N. Beckmann, C. Celio, J. Eastep, and A. Agarwal. Graphite: A distribuyted parallel simulator for multicores. In *Proceedings of the International Symposium on High Performance Computer Architecture (HPCA)*, pages 295–306, January 2010. 97, 100

[141] C. C. Minh, J. Chung, C. Kozyrakis, and K. Olukotun. STAMP: Stanford transactional applications for multi-processing. In *Proceedings of the IEEE International Symposium on Workload Characterization (IISWC)*, pages 35–46, September 2008. DOI: 10.1109/IISWC.2008.4636089 16

[142] S. S. Mukherjec, S. K. Reinhardt, B. Falsafi, M. Litzkow, M. D. Hill, D. A. Wood, S. Huss-Lederman, and J. R. Larus. Wisconsin wind tunnel II: A fast, portable parallel architecture simulator. *IEEE Concurrency*, 8(4):12–20, October 2000. DOI: 10.1109/4434.895100 51, 97, 100

[143] O. Mutlu, H. Kim, D. N. Armstrong, and Y. N. Patt. Understanding the effects of wrong-path memory references on processor performance. In *Proceedings of the 3rd Workshop on Memory Performance Issues (WMPI) held in conjunction with the 31st International Symposium on Computer Architecture (ISCA)*, pages 56–64, June 2005. 55

[144] S. Narayanasamy, C. Pereira, H. Patil, R. Cohn, and B. Calder. Automatic logging of operating system effects to guide application level architecture simulation. In *Proceedings of the ACM Sigmetrics International Conference on Measurement and Modeling of Computer Systems (SIGMETRICS)*, pages 216–227, June 2006. 53

[145] I. Nestorov, M. Rowland, S. T. Hadjitodorov, and I. Petrov. Empirical versus mechanistic modelling: Comparison of an artificial neural network to a mechanistically based model for quantitative structure pharmacokinetic relationships of a homologous series of barbiturates. *The AAPS Journal*, 1(4):5–13, December 1999. 32

[146] A.-T. Nguyen, P. Bose, K. Ekanadham, A. Nanda, and M. Michael. Accuracy and speed-up of parallel trace-driven architectural simulation. In *Proceedings of the 11th International Parallel Processing Symposium (IPPS)*, pages 39–44, April 1997. DOI: 10.1109/IPPS.1997.580842 95

[147] A. Nohl, G. Braun, O. Schliebusch, R. Leupers, and H. Meyr. A universal technique for fast and flexible instruction-set architecture simulation. In *Proceedings of the 39th Design Automation Conference (DAC)*, pages 22–27, June 2002. DOI: 10.1145/513918.513927 72

[148] D. B. Noonburg and J. P. Shen. A framework for statistical modeling of super-scalar processor performance. In *Proceedings of the Third International Symposium on High-Performance Computer Architecture (HPCA)*, pages 298–309, February 1997. DOI: 10.1109/HPCA.1997.569691 92

[149] S. Nussbaum and J. E. Smith. Modeling superscalar processors via statistical simulation. In *Proceedings of the 2001 International Conference on Parallel Architectures and Compilation Techniques (PACT)*, pages 15–24, September 2001. DOI: 10.1109/PACT.2001.953284 81, 86, 88

[150] S. Nussbaum and J. E. Smith. Statistical simulation of symmetric multiprocessor systems. In *Proceedings of the 35th Annual Simulation Symposium 2002*, pages 89–97, April 2002. DOI: 10.1109/SIMSYM.2002.1000093 91

[151] K. Olukotun, B. A. Nayfeh, L. Hammond, K. Wilson, and K.-Y. Chang. The case for a single-chip multiprocessor. In *Proceedings of the International Conference on Architectural Support for Programming Languages and Operating Systems (ASPLOS)*, pages 2–11, October 1996. 7

[152] M. Oskin, F. T. Chong, and M. Farrens. HLS: Combining statistical and symbolic simulation to guide microprocessor design. In *Proceedings of the 27th Annual International Symposium*

on Computer Architecture (ISCA), pages 71–82, June 2000. DOI: 10.1145/339647.339656 81, 83, 86, 88

[153] H. Patil, R. Cohn, M. Charney, R. Kapoor, A. Sun, and A. Karunanidhi. Pinpointing representative portions of large Intel Itanium programs with dynamic instrumentation. In *Proceedings of the 37th Annual International Symposium on Microarchitecture (MICRO)*, pages 81–93, December 2004. 70, 74

[154] M. Pellauer, M. Vijayaraghavan, M. Adler, Arvind, and J. S. Emer.   Quick performance models quickly: Closely-coupled partitioned simulation on FPGAs. In *Proceedings of the IEEE International Symposium on Performance Analysis of Systems and Software (ISPASS)*, pages 1–10, April 2008. DOI: 10.1109/ISPASS.2008.4510733 102

[155] D. A. Penry, D. Fay, D. Hodgdon, R. Wells, G. Schelle, D. I. August, and D. Connors. Exploiting parallelism and structure to accelerate the simulation of chip multi-processors. In *Proceedings of the Twelfth International Symposium on High Performance Computer Architecture (HPCA)*, pages 27–38, February 2006. 97, 102

[156] C. Pereira, H. Patil, and B. Calder.   Reproducible simulation of multi-threaded workloads for architecture design space exploration. In *Proceedings of the IEEE International Symposium on Workload Characterization (IISWC)*, pages 173–182, September 2008. DOI: 10.1109/IISWC.2008.4636102 59

[157] E. Perelman, G. Hamerly, and B. Calder. Picking statistically valid and early simulation points. In *Proceedings of the 12th International Conference on Parallel Architectures and Compilation Techniques (PACT)*, pages 244–256, September 2003. 68

[158] E. Perelman, J. Lau, H. Patil, A. Jaleel, G. Hamerly, and B. Calder. Cross binary simulation points. In *Proceedings of the Annual International Symposium on Performance Analysis of Systems and Software (ISPASS)*, March 2007. 70

[159] D. G. Perez, G. Mouchard, and O. Temam. MicroLib: A case for the quantitative comparison of micro-architecture mechanisms. In *Proceedings of the 37th Annual International Symposium on Microarchitecture (MICRO)*, pages 43–54, December 2004. 61

[160] A. Phansalkar, A. Joshi, and L. K. John. Analysis of redundancy and application balance in the SPEC CPU2006 benchmark suite. In *Proceedings of the Annual International Symposium on Computer Architecture (ISCA)*, pages 412–423, June 2007. 21, 25, 27, 62, 103, 104

[161] J. Rattner. Electronics in the internet age. Keynote at the International Conference on Parallel Architectures and Compilation Techniques (PACT), September 2001. 5

[162] J. Reilly. Evolve or die: Making SPECâŁ™s CPU suite relevant today and tomorrow. IEEE International Symposium on Workload Characterization (IISWC), October 2006. Invited presentation. DOI: 10.1109/IISWC.2006.302735 21

[163] S. K. Reinhardt, M. D. Hill, J. R. Larus, A. R. Lebeck, J. C. Lewis, and D. A. Wood. The wisconsin wind tunnel: Virtual prototyping of parallel computers. In *Proceedings of the ACM SIGMETRICS Conference on Measurement and Modeling of Computer Systems*, pages 48–60, May 1993. DOI: 10.1145/166955.166979 71, 97, 100

[164] M. Reshadi, P. Mishra, and N. D. Dutt. Instruction set compiled simulation: a technique for fast and flexible instruction set simulation. In *Proceedings of the 40th Design Automation Conference (DAC)*, pages 758–763, June 2003. DOI: 10.1145/775832.776026 72

[165] J. Ringenberg, C. Pelosi, D. Oehmke, and T. Mudge. Intrinsic checkpointing: A methodology for decreasing simulation time through binary modification. In *Proceedings of the IEEE International Symposium on Performance Analysis of Systems and Software (ISPASS)*, pages 78–88, March 2005. DOI: 10.1109/ISPASS.2005.1430561 73

[166] E. M. Riseman and C. C. Foster. The inhibition of potential parallelism by conditional jumps. *IEEE Transactions on Computers*, C-21(12):1405–1411, December 1972. DOI: 10.1109/T-C.1972.223514 35

[167] M. Rosenblum, E. Bugnion, S. Devine, and S. A. Herrod. Using the SimOS machine simulator to study complex computer systems. *ACM Transactions on Modeling and Computer Simulation (TOMACS)*, 7(1):78–103, January 1997. DOI: 10.1145/244804.244807 53

[168] E. Schnarr and J. R. Larus. Fast out-of-order processor simulation using memoization. In *Proceedings of the Eighth International Conference on Architectural Support for Programming Languages and Operating Systems (ASPLOS)*, pages 283–294, October 1998. DOI: 10.1145/291069.291063 71

[169] J. P. Shen and M. H. Lipasti. *Modern Processor Design: Fundamentals of Superscalar Processors*. McGraw-Hill, 2007. 5

[170] T. Sherwood, E. Perelman, and B. Calder. Basic block distribution analysis to find periodic behavior and simulation points in applications. In *Proceedings of the International Conference on Parallel Architectures and Compilation Techniques (PACT)*, pages 3–14, September 2001. DOI: 10.1109/PACT.2001.953283 68

[171] T. Sherwood, E. Perelman, G. Hamerly, and B. Calder. Automatically characterizing large scale program behavior. In *Proceedings of the International Conference on Architectural Support for Programming Languages and Operating Systems (ASPLOS)*, pages 45–57, October 2002. 68

[172] K. Skadron, P. S. Ahuja, M. Martonosi, and D. W. Clark. Branch prediction, instruction-window size, and cache size: Performance tradeoffs and simulation techniques. *IEEE Transactions on Computers*, 48(11):1260–1281, November 1999. DOI: 10.1109/12.811115 67

[173] J. E. Smith. Characterizing computer performance with a single number. *Communications of the ACM*, 31(10):1202–1206, October 1988. DOI: 10.1145/63039.63043 11

[174] A. Snavely and D. M. Tullsen. Symbiotic jobscheduling for simultaneous multithreading processor. In *Proceedings of the International Conference on Architectural Support for Programming Languages and Operating Systems (ASPLOS)*, pages 234–244, November 2000. 9, 10

[175] D. J. Sorin, V. S. Pai, S. V. Adve, M. K. Vernon, and D. A. Wood. Analytic evaluation of shared-memory systems with ILP processors. In *Proceedings of the 25th Annual International Symposium on Computer Architecture (ISCA)*, pages 380–391, June 1998. 46

[176] A. Srivastava and A. Eustace. ATOM: A system for building customized program analysis tools. Technical Report 94/2, Western Research Lab, Compaq, March 1994. 22, 51

[177] R. A. Sugumar and S. G. Abraham. Efficient simulation of caches under optimal replacement with applications to miss characterization. In *Proceedings of the 1993 ACM Conference on Measurement and Modeling of Computer Systems (SIGMETRICS)*, pages 24–35, 1993. DOI: 10.1145/166955.166974 54

[178] D. Sunwoo, J. Kim, and D. Chiou. QUICK: A flexible full-system functional model. In *Proceedings of the Annual International Symposium on Performance Analysis of Systems and Software (ISPASS)*, pages 249–258, April 2009. 57, 102

[179] P. K. Szwed, D. Marques, R. B. Buels, S. A. McKee, and M. Schulz. SimSnap: Fast-forwarding via native execution and application-level checkpointing. In *Proceedings of the Workshop on the Interaction between Compilers and Computer Architectures (INTERACT), held in conjunction with HPCA*, February 2004. DOI: 10.1109/INTERA.2004.1299511 71

[180] T. M. Taha and D. S. Wills. An instruction throughput model of superscalar processors. *IEEE Transactions on Computers*, 57(3):389–403, March 2008. DOI: 10.1109/TC.2007.70817 45

[181] N. Tuck and D. M. Tullsen. Initial observations of the simultaneous multithreading Pentium 4 processor. In *Proceedings of the International Conference on Parallel Architectures and Compilation Techniques (PACT)*, pages 26–34, September 2003. 7

[182] D. M. Tullsen, S. J. Eggers, and H. M. Levy. Simultaneous multithreading: Maximizing on-chip parallelism. In *Proceedings of the 22nd Annual International Symposium on Computer Architecture (ISCA)*, pages 392–403, June 1995. DOI: 10.1109/ISCA.1995.524578 7

[183] M. Vachharajani, N. Vachharajani, D. A. Penry, J. A. Blome, and D. I. August. Microarchitectural exploration with Liberty. In *Proceedings of the 35th International Symposium on Microarchitecture (MICRO)*, pages 271–282, November 2002. 61

[184] M. Van Biesbrouck, B. Calder, and L. Eeckhout. Efficient sampling startup for SimPoint. *IEEE Micro*, 26(4):32–42, July 2006. DOI: 10.1109/MM.2006.68 73

[185] M. Van Biesbrouck, L. Eeckhout, and B. Calder. Efficient sampling startup for sampled processor simulation. In *2005 International Conference on High Performance Embedded Architectures and Compilation (HiPEAC)*, pages 47–67, November 2005. DOI: 10.1007/11587514_5 76

[186] M. Van Biesbrouck, L. Eeckhout, and B. Calder. Considering all starting points for simultaneous multithreading simulation. In *Proceedings of the International Symposium on Performance Analysis of Systems and Software (ISPASS)*, pages 143–153, March 2006. 78

[187] M. Van Biesbrouck, L. Eeckhout, and B. Calder. Representative multiprogram workloads for multithreaded processor simulation. In *Proceedings of the IEEE International Symposium on Workload Characterization (IISWC)*, pages 193–203, October 2007. DOI: 10.1109/IISWC.2007.4362195 78

[188] M. Van Biesbrouck, T. Sherwood, and B. Calder. A co-phase matrix to guide simultaneous multithreading simulation. In *Proceedings of the International Symposium on Performance Analysis of Systems and Software (ISPASS)*, pages 45–56, March 2004. DOI: 10.1109/ISPASS.2004.1291355 78, 91

[189] T. F. Wenisch, R. E. Wunderlich, B. Falsafi, and J. C. Hoe. Simulation sampling with livepoints. In *Proceedings of the Annual International Symposium on Performance Analysis of Systems and Software (ISPASS)*, pages 2–12, March 2006. 73, 76

[190] T. F. Wenisch, R. E. Wunderlich, M. Ferdman, A. Ailamaki, B. Falsafi, and J. C. Hoe. SimFlex: Statistical sampling of computer system simulation. *IEEE Micro*, 26(4):18–31, July 2006. DOI: 10.1109/MM.2006.79 7, 55, 66, 78, 79, 95

[191] E. Witchell and M. Rosenblum. Embra: Fast and flexible machine simulation. In *Proceedings of the ACM SIGMETRICS Conference on Measurement and Modeling of Computer Systems*, pages 68–79, June 1996. 51, 54, 71

[192] D. A. Wood, M. D. Hill, and R. E. Kessler. A model for estimating trace-sample miss ratios. In *Proceedings of the 1991 SIGMETRICS Conference on Measurement and Modeling of Computer Systems*, pages 79–89, May 1991. DOI: 10.1145/107971.107981 75

[193] R. E. Wunderlich, T. F. Wenisch, B. Falsafi, and J. C. Hoe. SMARTS: Accelerating microarchitecture simulation via rigorous statistical sampling. In *Proceedings of the Annual International Symposium on Computer Architecture (ISCA)*, pages 84–95, June 2003. DOI: 10.1145/859618.859629 64, 66, 67, 74, 77, 93

[194] R. E. Wunderlich, T. F. Wenisch, B. Falsafi, and J. C. Hoe. Statistical sampling of microarchitecture simulation. *ACM Transactions on Modeling and Computer Simulation*, 16(3):197–224, July 2006. DOI: 10.1145/1147224.1147225 64, 66, 67, 74, 77, 93

[195] J. J. Yi, S. V. Kodakara, R. Sendag, D. J. Lilja, and D. M. Hawkins. Characterizing and comparing prevailing simulation techniques. In *Proceedings of the International Symposium on High-Performance Computer Architecture (HPCA)*, pages 266–277, February 2005. 70

[196] J. J. Yi, D. J. Lilja, and D. M. Hawkins. A statistically rigorous approach for improving simulation methodology. In *Proceedings of the Ninth International Symposium on High Performance Computer Architecture (HPCA)*, pages 281–291, February 2003. DOI: 10.1109/HPCA.2003.1183546 27, 34

[197] J. J. Yi, R. Sendag, L. Eeckhout, A. Joshi, D. J. Lilja, and L. K. John. Evaluating benchmark subsetting approaches. In *Proceedings of the 2006 IEEE International Symposium on Workload Characterization (IISWC)*, pages 93–104, October 2006. DOI: 10.1109/IISWC.2006.302733 29

[198] J. J. Yi, H. Vandierendonck, L. Eeckhout, and D. J. Lilja. The exigency of benchmark and compiler drift: Designing tomorrow's processors with yesterday's tools. In *Proceedings of the 20th ACM International Conference on Supercomputing (ICS)*, pages 75–86, June 2006. DOI: 10.1145/1183401.1183414 17

[199] M. T. Yourst. PTLsim: A cycle accurate full system x86-64 microarchitectural simulator. In *Proceedings of the International Symposium on Performance Analysis of Systems and Software (ISPASS)*, pages 23–34, April 2007. 55, 61

Author's Biography

LIEVEN EECKHOUT

Lieven Eeckhout is an Associate Professor at Ghent University, Belgium. His main research interests include computer architecture and the hardware/software interface in general, and performance modeling and analysis, simulation methodology, and workload characterization in particular. His work was awarded twice as an IEEE Micro Top Pick in 2007 and 2010 as one of the previous year's "most significant research publications in computer architecture based on novelty, industry relevance and long-term impact". He has served on a couple dozen program committees, he was the program chair for ISPASS 2009 and general chair for ISPASS 2010, and he serves as an associate editor for ACM Transactions on Architecture and Code Optimization. He obtained his Master's degree and Ph.D. degree in computer science and engineering from Ghent University in 1998 and 2002, respectively.

Printed in the United States
by Baker & Taylor Publisher Services